PLEASE STAMP DATE DUE, BOTH BELOW AND ON CARD

DATE DUE	DATE DUE	DATE DUE	DATE DUE

GL-15

Workshop on Databases for Galactic Structure

Sponsored by

Swarthmore College
University of Lausanne
Institute for Space Observations

May 17 - 19, 1993

Meeting held at Swarthmore College, Swarthmore, Pennsylvania

Edited by

A. G. Davis Philip
Union College and Van Vleck Observatory

Bernard Hauck
University of Lausanne

and

Arthur R. Upgren
Van Vleck Observatory

Van Vleck Observatory Contribution No. 13

L. Davis Press
Schenectady, N. Y.
1993

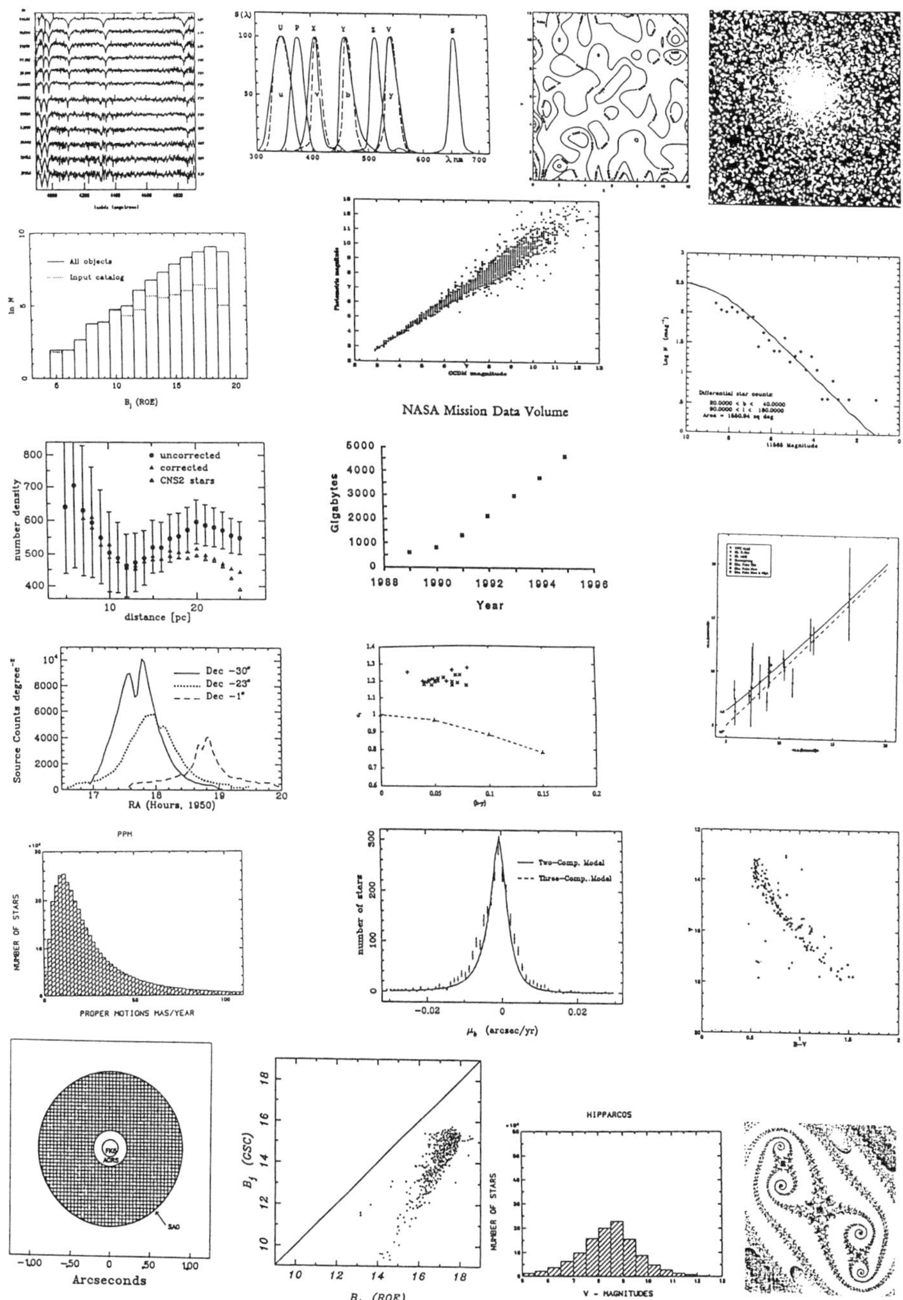

Wilhelm Gliese

This meeting is dedicated to the memory of Wilhelm Gliese, who died on June 12, 1993, shortly after the meeting was held. Wilhelm Gliese's name has become rightly synonymous with the study of the nearby stars. In 1957, and again in 1969, he provided a database on all stars likely to lie within 20 and in the later edition 22 parsecs of the Solar System. These two editions of his Catalogue of Nearby Stars represented the best definition of the galactic disk population available. In 1979 he and Hartmut Jahreiss added a supplement to the second edition. At the time of his death Gliese had just completed his part of the third edition, soon to be published.

The care and concern which Dr. Gliese took to assure the best data possible were included in his catalogues, are models of the effort to be made in assembling a basic body of observational data. Although the coming third edition, which includes some 3800 stars to a limiting distance of 25 parsecs, serves as a memorial to him and his dedication, we wish to honor him here as well. Hence this volume is dedicated to his memory.

Wilhelm Gliese

Table of Contents

Section II - Poster Papers

PREFACE

This meeting on Databases for Galactic Structure was the idea of Bernard Hauck. The technology of data collection and storage has been evolving rapidly, enabling ever larger collections of data to be made and then be used by the astronomical community. Some astronomical databases now include tens of terabytes and soon we will probably reach the level of petabyte databases. How best to manage and distribute the information contained in these large bodies of data are important topics for discussion. At the workshop speakers described many projects involving the collection of large amounts of information (from Starbase which will contain data on nearly a billion stellar objects and several million galaxies to smaller projects involving the study of objects in a small region of the Milky Way). There were two discussion sessions; one on "How can we Improve Access to Data and Data Centers" and another on "Which Kinds of Surveys will be Most Useful for Those Who Build Galactic Models?". Barry Lasker and Alain Omont were the discussion leaders.

The question of where to hold the meeting was solved when Wulff Heintz offered to sponsor the meeting at Swarthmore College in May, 1993. Forty-one astronomers from eight countries attended. The oral papers were allotted 30 minutes which gave ample time for each topic. The poster papers were displayed on boards inside the meeting room and were up for the three days of the meeting. They were discussed in a discussion period on the third day.

The Scientific Organizing Committee was formed of the following astronomers:

Bernard Hauck	Lausanne University	Chair
Michel Crézé	Observatory of Strasbourg	
Wulff Heintz	Swarthmore College	
Roberta Humphreys	University of Minnesota	
Jaylee Mead	NASA\Goddard Space Flight Center	
A. G. Davis Philip	Union College and VVO	
William van Altena	Yale University Observatory	

The Local Organizing Committee was formed of the following astronomers:

Wulff Heintz	Swarthmore College	Chair
Harry Augenson	Swarthmore College	
A. G. Davis Philip	Union College and VVO	

Special thanks are due to Harry Augenson and Bruce Cantor who shared the job of typing up the discussion sheets following each paper and then distributing them back to each person involved for proofreading. Most of the discussion was in good shape before the meeting ended. The final details were finished on email.

Papers were submitted via email or on floppy disks, with a hardcopy version mailed to Schenectady. When the papers were printed out on the laser printer copies were sent to each author for proofreading. Each of the three editors also read all the manuscript. WordPerfect 5.1 was used to create the final manuscript.

On the back cover of this book a diagram of the nearby stars (from Luyten 1926, The Scientific Monthly 22, 494) is reprinted. If you look at the two pictures and cross your eyes slightly you will see three square areas. Focus on the middle square and the stars will appear in 3-D. The model represents the stars within 10 pc of the Sun, as seen from a point about 50 pc in the direction of the NGP.

Fairbanks
December, 1993

A. G. Davis Philip

List of Participants

France

Michel Crézé	Observatory of Strasbourg
Nicolas Epchtein	Paris Observatory
Edouard Oblak	Besançon Observatory
Alain Omont	

Germany

Hartmut Jahreiss	Astronomisches Rechen Institut
Siegfried Roeser	Astronomisches Rechen Institut

Spain

Peter Hammersley	Tenerife

Lithuania

Vytas Straižys	Inst. for Theo. Phys. and Astron.

Switzerland

Bernard Hauck	University of Lausanne
Jean-Claude Mermilliod	University of Lausanne

Taiwan

Wean Shun Tsay	National Central University

UK

Elizabeth Griffin	Institute for Astronomy, Cambridge

USA

Tom Bania	Boston University
Richard P. Boyle	Vatican Obs., Univ. of Arizona
Stefano Casertano	University of Illinois
Martin Cohen	University of California, Berkeley
Guenther Eichhorn	Smithsonian Astrophysical Obs.
Michael Garcia	Center for Astrophysics
Terry Girard	Yale University
Xinjian Guo	Yale University
Wulff Heintz	Swarthmore College

Roberta Humphreys — University of Minnesota
Barry M. Lasker — Space Telescope Science Inst.
Mario Lattanzi — Space Telescope Science Inst.
John T. Lee — Yale University
Felix J. Lockman — NRAO, W. Virginia
Wenzhang Ma — Yale University
Rene A. Méndez — Yale University
Phillip Lu — Western Connecticut State Univ.
A. G. Davis Philip — Union College and VVO
Jeffrey Pier — US Naval Observatory
Imants Platais — Yale University
Vera Platais — Yale University
Kavan Ratnatunga — Johns Hopkins University
Michael Richmond — Princeton University
Nancy G. Roman — NASA/Goddard Space Flight Center
Arthur R. Upgren — Van Vleck Observatory
Sean Urban — US Naval Observatory
Wayne Warren — NASA/Goddard Space Flight Center
Charles Worley — US Naval Observatory
William van Altena — Yale University

Review Papers

Chairs of Sessions

I	Roberta Humphreys
II	A. G. Davis Philip
III	Michel Crézé
IV	W. F. van Altena
V	Arthur R. Upgren
VI	Wulff Heintz

DATABASES: PAST AND PRESENT

B. Hauck

Institute of Astronomy of the University of Lausanne

Data collection played a very important role in astronomy as far back as the second century B.C. It was the epoch of the first catalogue of stars, a catalogue, as you know, made by Hipparchus concerning 1080 stars. A very short list by today's standards! The main reason for this importance is undoubtedly that we have no direct access to the objects we are studying and no possibility of manipulating them. If data are archived, and of course if we have access to them, we can avoid performing new observations of the same objects with the same kind of receptors. We can also look for variability of the objects. The first way of gathering astronomical data was to compile a catalogue and then to distribute it. That method is still in use today. But catalogues are the reflection of technical development. Bigger telescopes enable us to observe more stars, thus the size of the catalogues increases! New detectors are used and new catalogues are published. Table 2.4 of Mihalas (1968) is a good illustration of that in the visible and the situation in the other wavelength domains also confirms it.

New techniques for storing information have been used for astronomical catalogues since the sixties: punched cards and magnetic tapes. The concept of a computerised astronomical data center appeared. We find in the literature at least three proposals (Vallee and Hynek 1966, Bidelman 1967, Jaschek 1968). That epoch also saw the flowering of various small "data centers". A quick look at the CODATA Directory of data sources for sciences and technology (C. Jaschek 1980) shows seventy data centers! I will not discuss here the definition of a data center used by Jaschek. The large number of "data centers" is only an illustration of the interest for the data and also an illustration of the need to gather together the data concerning one field.

It was also the epoch of two projects for large data centers. At its May 2, 1969 meeting the Board of Directors of the Astronomical Society of the Pacific established an Ad Hoc committee on the feasibility of an astronomical data center. Dr H. Abt chaired this committee, whose report mentions that early attempts to create data centers (of course before the availability of electronic computers) were made by Lundmark, Annie Cannon and various Russian astronomers. We also find in this report a table with all catalogues on punched cards or magnetic tapes existing at that time. Fifteen catalogues are mentioned! The intention of this committee was to seek the collaboration of observatories which were active in "measuring, compiling, intercomparing and publishing data in certain branches of astronomy". At the same time, but independently, a French group headed by Professor Delhaye was establishing a project that called for the collaboration not only of French astronomers, but also of Swiss and German astronomers from various observatories, each specializing in a specific domain and continuing to work in their "natural" scientific environment. It was in fact the early creation of a network

Workshop on Databases for Galactic Structure
A. G. Davis Philip, Bernard Hauck and Arthur
R. Upgren, eds. p. 3 - 8 © 1993 L. Davis Press

Bernard Hauck
Institute of Astronomy
University of Lausanne
Lausanne, Switzerland

In 1972 the Stellar Data Center of Strasbourg was created. Many catalogues were gathered at the CDS and a first list of those available, on magnetic tapes, was published in 1974 (Hauck and Jung 1974). 75 stellar catalogues were mentioned. Along with the collection of catalogues or the encouragement to create new catalogues, the necessity to create a database became obvious. The purpose of this database was not only to include all data gathered at Strasbourg but also to allow cross-identification of stars. Thus SIMBAD (Set of Identifications, Measurements and Bibliography for Astronomical Data) was born. That was a very pioneering enterprise! Twenty years later the situation is quite different. SIMBAD contains 740,000 objects (640,000 stellar and 100,000 non-stellar). Egret et al. (1991) have given a very good review of the present contents and possibilities of this database. Furthermore, SIMBAD is no longer the only database and the CDS is no longer the only data center! During the last twenty years we have witnessed an explosion of new data together with an explosion of technical support. However the data explosion follows the same pattern as in the past. New detectors in new spectral domains and new databases are created or a new large project is launched and a database is foreseen. This is well illustrated in a recent book by M. A. Albrecht and D. Egret (1991). Several chapters in this book concern archives or databases for the following instruments: ROSAT, EXOSAT, IRAS, COBE, IUE, HST and HIPPARCOS. Other chapters are devoted to SIMBAD and the NASA/IPAC Extragalactic Databases. These are not the only changes that have occurred. Up to a few years ago the consulting of catalogues was a laborious task. Nowadays, seated at your workstation, you can consult many databases that are perhaps located in various countries. In addition, the concept of a catalogue has changed.

Catalogues used to be printed as an end product and many years would elapse before the appearance of a hypothetical updated edition. Catalogues and databases are evolving and this leads to a need for remote access. You can be connected through a wide area network such as Internet or Bitnet as well as through a public data transmission network like Transpac in France, Telenet in the USA, etc. Two networks are specially dedicated to astronomy: Starlink (Davenhall 1991) and Astronet (Benacchio 1991). In fact both are more than a network. A computer network is the hardware, while the software consists of a collection of different kinds of software to reduce and analyze astronomical data. In both cases, database software has developed.

If remote access is impossible or expensive you can use CD-ROMs, which are becoming increasingly important. CD-ROMs are a very cheap way to distribute data to various users. We can assume that CD-ROMs will be used not only to distribute catalogues but also primary data, as the Astrophysical Journal has the intention to do.

The discussion above has illustrated the evolution from catalogues in machine-readable form to databases. It seems to me we can follow Heck (1992) in his definition of a database: "The term database could be generic and include archives as well. A stricter meaning however would link it to a context of the scientific data extracted or derived from observational material after reduction and/or analysis, although a database can also be made of non-observational data. Contrary to a databank, a database would also generally contain some software to retrieve the data and possibly to work on them".

The transportation of data is another major change that should also be mentioned. Originally a standard format for transport of astronomical images, FITS (Flexible Image

Transport System) has been developed since 1979. At the General Assembly of the IAU in Patras in 1982, the use of FITS was recommended by Commission 5. During a meeting of this commission, the extension of the FITS format to table and catalogue data was discussed. This format is now widely used and most of the major observatories have introduced it.

As we have seen above, space astronomy has generated large databases, while the situation with groundbased data has remained somewhat anarchic. The only effort to collect large samples of data is that of the CDS and its associated observatories. However, the situation is perhaps now changing. Some observatories (Ochsenbein 1991, Raimond 1991, Wells 1991) have started the organization of archives at ESO, La Palma and VLA respectively.

A very good review of this aspect of the collection of data is given by M. A. Albrecht (1993). Furthermore, an ESO-OAT international workshop was held last month in Trieste on this topic. The number of observatories taking care of the archiving of their observations is growing and we may wonder whether data centers will still be useful! This problem was discussed at the last meeting of the CDS in September 1992. I extract the following passage from the concluding remarks made by Crézé and Pasian (1993): "The role of data centers (e.g. CDS, NSSDC) in this framework is somewhat different: it aims at the harmonization of efforts and at the archiving of more elaborate collections of data collected downstream, in connection with the process of publication. Data centers should also care for multiwavelength and multitechnic compilations. Suggested actions are: a) to foster direct collaboration with projects, aimed at transferring the scientific expertise necessary to understand the archived data after the possible dispersion of some of the project teams; b) to define with the observatories which is the subset of the information contained in the observation catalogues which is useful for archival researchers; c) to collect and make available via network observation catalogues and general information on available data; d) to collect and make available via network homogeneous data (results from surveys, etc.); e) to coordinate the implementation of standard descriptors which would serve both for the publication of data and their automated insertion in on-line databases."

Data centers still have a role to play! In fact this role will become more and more important. Twenty years ago an aim of the CDS was to gather in one location all stellar data to facilitate their use. That is now the case for the majority of data obtained in the visible region of the spectrum. However as we have seen above, there are many other databases in various spectral domains. Information concerning astronomical data is again dispersed but now it is in many databases! To remedy this situation, both ESA and NASA have started projects on an information system. M. A. Albrecht (1991) has reviewed the aims of ESIS (European Space Information System). ESIS will devise a set of tools to access, exchange and handle information from space missions, archives, databases, etc. NASA is developing the "Astrophysics Data System" (Giovane 1992, Weiss and Good 1991). The ADS "provides User Services for location retrieval, and some analysis of on-line data from space-based NASA missions, and for retrieval of documentation relevant to the data and to astrophysical research in general" (Weiss and Good). The ADS structure is a very interesting one. It consists of primary nodes responsible for providing astrophysical data in their field, for example a space mission. Each node offers services specific to its database. The ADS provides the user with an interface to this network of databases.

Archiving astronomical data plays an important role today and much is being done in this field. Concerning the space-based data, the situation is presently satisfactory. However a great deal of data are still scattered throughout the astronomical literature and they will be lost if they are not included in a database. We have to encourage all efforts to compile data and to conserve them.

My last remark is taken from Benvenuti (1992). In his review at the Conference on Astronomy from Large Databases II, he quoted that at the first conference he had noticed that the percentage of papers dedicated to an astronomical application was very low and unfortunately he could make the same statement at the second. He then continued: "My personal opinion is that archives and databases are today an extremely useful (and used) tool in the astronomical and astrophysical research, but, with exceptions, they have not yet become an alternative source of data with respect to new observations." I think that Benvenuti is only partly right. Many people are using data, even archived data, but very few are interested in the management of databases. They want to use them as an end product, which is the reason why workshops and conferences on this kind of topic attract more "database makers" than "data users". The participation in this workshop is also a good illustration.

REFERENCES

Albrecht, M. A. 1991 in Databases and On-line Data in Astronomy, M. A. Albrecht and D. Egret, eds., Kluwer Acad. Publ., Dordrecht, p. 127

Albrecht, M. A. 1993 in Astronomical Data Analysis Software and systems, in press

Albrecht, M. A. and Egret, D. eds. 1991 Databases and On-line Data in Astronomy, Kluwer Acad. Publ., Dordrecht

Benacchio, L. 1991 in Databases and On-line Data in Astronomy, M. A. Albrecht and D. Egret, eds., Kluwer Acad. Publ., Dordrecht, p. 179

Benvenuti, P. 1991 in Astronomy from Large Databases II, A. Heck, and F. Murtagh, ESO Conference and Workshop Proc., 43, p. 43

Bidelman, W. P. 1967 Trans. IAU 13A, 1017

Crézé, M. and Pasian, F. 1993 Information Bull. CDS 45, in press

Davenhall, C. 1991 in Databases and On-line Data in Astronomy, M. A. Albrecht and D. Egret, eds., Kluwer Acad. Publ., Dordrecht, p. 167

Egret, D., Wenger, M. and Dubois, P. 1991 in Databases and On-line Data in Astronomy, M. A. Albrecht and D. Egret, Kluwer Acad. Publ., Dordrecht, p. 79

Giovane, F. 1992 in Data Analysis in Astronomy IV, V. Di Gesu et al., eds., Plenum Press, N. Y., p. 3

Hauck, B. and Jung, J. 1972 A&AS 16, 289

Heck, A. 1992 in Data Analysis in Astronomy IV, V. Di Gesu et al. (eds.), Plenum Press, N.Y., p. 21

Jaschek, C. 1968 PASP 80, 654

Jaschek, C. 1980 CODATA Bull. 36, 1

Mihalas, D 1968 Galactic Astronomy, W.H. Freeman, San Francisco

Ochsenbein, F. 1991 in Databases and On-line Data in Astronomy, M. A. Albrecht and D. Egret, eds., Kluwer Acad. Publ., Dordrecht, p. 107

Raimond, E. 1991 in Databases and On-line Data in Astronomy, M. A. Albrecht and D. Egret, eds., Kluwer Acad. Publ., Dordrecht, p. 115

Vallee, J. F. and Hynek, J. A. 1966 PASP 78, 315
Weiss, J. R. and Good, J. C. 1991 in Databases and On-line Data in Astronomy, M. A. Albrecht and D. Egret, eds., Kluwer Acad. Publ., Dordrecht, p. 139
Wells, D. C. 1991 in Databases and On-line Data in Astronomy, M. A. Albrecht and D. Egret, eds., Kluwer Acad. Publ., Dordrecht, p. 125

8   B. Hauck

DISCUSSION

GARCIA: How has SIMBAD evolved over the last 20 years?

HAUCK: The Director of the CDS, Dr. Crézé, is in the audience and he can answer better than I!

CREZE: The SIMBAD database was created in 1976. First it consisted of cross-identifications of a few main astrometric and photometric catalogues. It was conceived as a tool for stellar statistics in galactic structure problems. Later it was merged with a bibliographic database and then it was permanently upgraded to a bibliographic survey and it is now used in nearly all fields of astronomy. The database management system was redesigned in 1989. After the recent cross-identifications with the IRAS Point Source Catalogue, Simbad contains currently more than one million objects. Another inovation is the "question" Email box where all requests for corrections or information on Simbad usage are answered by the CDS staff within 24 hours. In 1986 some 250 users opened less than 1000 sessions monthly; those figures evolved to 1000 and more than 4000 currently.

GRIFFIN: Despite the recognized potential of databases, sympathy - and therefore funding - for their handling and distribution has an unfortunate dependence upon national policies and national attitudes for funding scientific research. In the UK, for example, sponsors of research are intent on purchasing RESULTS, leaving the creation of links between the database(s) and the user(s) to the initiative of (a) volunteer(s). That sort of situation should be a matter of serious concern at this meeting.

PHILIP: Are any of the data centers collecting CCD data? Now, if we wish to find out information concerning celestial objects 100 years ago we can go to the Harvard Plate Archives, Warner & Swasey Observatory plate vault and other plate collections and find the information we need. But if we imagine people 100 years from now looking back at our time there will be no record of the sky available. CCD data are scattered all over and sometimes are overwritten by the next observer. A few places can archive CCD frames. I know that the Dominion Astrophysical Observatory now is archiving CCD frames taken with their telescopes. But at NOAO data are kept for only 6 months and then the tapes are reused.

HAUCK: To my knowledge, no data centers are collecting CCD data today.

HUMPHREYS: A comment on data archiving. During the past decade astronomers have been losing their record of the night sky with the increased and now almost exclusive use of CCDs for imaging (and spectra). These images have not been archived. There was a recommendation in support of data archiving in the Bahcall commission report. Initially, the major U.S. observatories have been reluctant to archive digital data because they are concerned about the extra cost. Even if they do eventually archive the observations, they will not want to be in the role of data distribution centers. The funding agencies will have to make the funds available if archiving is to succeed.

CD-ROMs IN (AND OUT OF) ASTRONOMY

Michael R. Garcia

Smithsonian Astrophysical Observatory, Center for Astrophysics

ABSTRACT: The electronic distribution of large astronomical databases has increased at an explosive pace during the last decade. This distribution has occurred via a combination of on-line access systems, magnetic tape distribution, and optical disk distribution. Examples of these methods, and the advantages of each, will be presented. Emphasis will be on CD-ROM as a distribution and archival medium and on the lessons learned with the EINSTEIN CD-ROM Archive. CD-ROM has cost/performance advantages that are presently unmatched by other distribution and archival media. Some interesting trends in CD-ROM technology development will be briefly mentioned.

1. REQUIREMENTS FOR ARCHIVE MEDIA

In planning out an astronomical data archive, one is faced with a series of cost/performance tradeoffs. These are driven by the size of the archive, the available budget, and the anticipated usage of the archive. The wide variation of these factors results in a wide range of archiving solutions. However, some requirements which most archives share are:

• Long Media Life: Meaning > 10 years. This requirement eliminates most magnetic tape, but is easily met by most optical storage media, including WORMs, Magneto-Optical, CD-ROM, (and photographic plates).

• Long Technology Life: If you cannot read the data off the media, the fact that the media has a long life is small consolation. Due to rapid improvements in computer and optical storage technology, the lifetime of optical technologies is notoriously short. A common saying in the computer industry is "10 years is forever". While one can always "archive" the reader with the media, the cost of maintaining and operating a specialized optical disk reader may be high.

• Large Storage Capacity: Astronomical archives of ~ 10 Terabytes are becoming common - several of them are being discussed in these proceedings (HST, Sturch 1993; Sloan Digital Sky Survey, Richmond 1993; DENIS, Epchtein 1993; StarBase, Humphreys 1993, etc.). With the storage capacity of optical disks (and mag tapes) in the ~ Gigabyte range, those needing to keep the archive "on-line" must use large juke-boxes.

• Low Cost! This is of course a prime consideration. One must consider the cost of the media itself (which can easily dominate) as well as the cost of readers, writers, and juke-boxes if needed.

• Interoperability: Archives may be held at a single (or a few) locations, and accessed via

Workshop on Databases for Galactic Structure
A. G. Davis Philip, Bernard Hauck and Arthur
R. Upgren, eds. p. 9 - 17 © 1993 L. Davis Press

Michael Garcia
Center for Astrophysics
60 Garden Street
Cambridge, Massachusetts
02138

the networks. Recently there has been a trend towards physically distributing all or selected portions of the archive. If the archive is to be distributed, there is an additional media requirement: Interoperability. As not all users have the same computer equipment or operating system, it may be necessary to make multiple, operating system specific, copies of the archive to distribute. Alternatively, one can look to a media/format that provides "interoperability" across computer platforms.

2. COST-PERFORMANCE STUDIES

Table 1 shows current costs for a variety of popular optical media and drives, and the equivalent cost/MB of storage capacity. In computing costs for CD-ROM, we have assumed

TABLE 1

Optical Media Costs

	Capacity	Media cost	Drive cost	Media cost/MB	Drive cost/MB
CD-ROM	0.66 G	$3.00	$500	0.005	0.76
CD/WO	0.66 G	$30.00	$500	0.050	0.76
5.25" M/O	0.60 G	$240.00	$5,000	0.040	8.30
5.25" WORM	0.6 0 G	$140.00	$5,000	0.230	8.30
12" WORM	1-8 G	$500.00	$7,000	0.060	0.88

TABLE 2

Optical Juke-Box Costs

	Capacity	Bytes	Drive cost	Loaded cost	Loaded cost/GB
CD-ROM	6 x 7	24GB	$1k x 7	$7k	0.29 $k/GB
CD/WO	240	160GB	$20k	$38k	0.16 $k/GB
5.25" M/O	30	18GB	$18k	$25k	1.40 $k/GB
5.25" M/O	160	100GB	$100k	$138k	1.40 $k/GB
12" WORM	100	800GB	$200k	$250k	0.30 $k/GB

it will be used to distribute 1000 copies of the archive (the number we found required for the EINSTEIN archive), that the first disk was $1000, and subsequent copies were $2. If ~ 100 copies are required the cost/MB increases by ~3x. From the table one sees that on a cost/MB basis CD-ROM is at least one order of magnitude cheaper than other media. If a single (or few) copies are needed, the new CD-ROM Write-Once (CD/WO) and WORM media and drives are competitively priced on a $/MB basis. Magneto/Optical media and drives, while an additional order of magnitude more expensive per MB, offer the ability to re-write to the same media.

In a similar fashion, Table 2 shows current costs for various optical juke-boxes and also the costs of "loading" the jukebox with media. Pioneer has developed a 6-slot mini-jukebox for CD-ROM, up to 7 of which can be daisy chained on a single computer SCSI bus. Several data centers maintain large sets of CD-ROMs on line using this method (einline.harvard.edu, explorer.arc.nasa.gov). A few large (100-240 disk) CD-ROM jukeboxes have recently become available. When filled with CD/WO disks, the 240 disk jukebox made by Kubik Enterprises (Martin 1993) has one of the lowest cost/byte figures available. Jukeboxes holding 12" WORM platters, like the ones used for the HST/DADS Archive, have a similar cost/byte figure but a large initial cost.

Some of the (dis)-advantages of various media which are harder to attach dollar figures to are missing from these two tables. In the case of CD-ROM, some of the advantages are: 1) the ISO-9660 format for CD-ROM gives it unique platform independence as it is supported by all major manufactures; 2) the five million CD-ROM drives installed on computers gives CD-ROM a VERY large "customer base"; 3) the conformance with CD-Audio and large customer base imply that CD-ROM/ISO9660 will be popular for a substantial time, while other optical disks formats (e.g., one gig LMSI 12" WORM) have already been dropped by manufacturers. Some of the disadvantages are: CD-ROMs only hold ~:seven gigabytes; the ~$1000 cost of a master disk creates a high threshold for cost-effective use of CD-ROMs (but CD/WO obviates this); there is no way to write incrementally to ISO9660/CD-ROM; and you cannot re-write or update CD-ROM, it is truly a write once only medium. This last item may not really be a disadvantage. In many applications, perhaps including archiving, write once is preferable to media which could be erased or over-written.

3. THE EINSTEIN EXPERIENCE

During the lifetime of the EINSTEIN Data Center at SAO many changes in the way in which NASA data centers operate have occurred. At the start of the project, in the late 70s, guest users came to the CfA in order to use the single centralized computer which ran EINSTEIN processing and analysis, and there was a limited ability for users to take data home on 9-track tape. At the close of the data center, in the early 1990s, the data and data products are accessible over the network via EINLINE and ADS, and are distributed on CD-ROMs. With CD-ROM distribution, the word "Data Center" may be a misnomer as the processed data need not be centralized, and analysis can be done using distributed software (IRAF/PROS). The total size of the processed EINSTEIN data archive is 40 Gigabytes. Distribution of the archive via CD-ROMs has allowed ~ four Terabytes of data to be distributed over the last three years. Logins to the EINLINE system continue to increase and currently run at ~300 /month.

TABLE 3

EINSTEIN CD-ROM Survey

Question	Yes	No	No Ans.	Other
Are EINSTEIN CD-ROM data useful in your work?	202	3	42	48
[Should] someone distribute other High Energy data products on CDROM?				
ROSAT	221	3	64	7
ASTRO-D	171	2	112	10
XTE	138	5	144	14

Many of the responses to question #1 tallied as "other" in Table 3 were "not yet", "Just starting", "I'm sure I will", etc.

Questions about the efficacy of distributing archives on CD-ROM led us to conduct an informal survey with the intent of discovering if the several thousand EINSTEIN CD-ROMs we have distributed are being used for archive data analysis (as intended), or as frissbees and coasters! The results (Table 3) were accumulated at AAS meetings and via a mailer sent out in the HEAO newsletter. They seem to indicate that CD-ROMs are useful as a data distribution tool, and that there is community interest in having other data sets similarly distributed.

In the case of EINSTEIN the data volume distributed via CD-ROM (four Terabytes/three years) is ~ 1000 times the volume distributed via EINLINE (the same data is available via both methods). However, data volume is only one measure of the efficacy of a data distribution effort, there are inherent (dis)-advantages to each method. From (perhaps) a users viewpoint, CD-ROM has advantages of: 1) rapid access, since the archive is "on your desk", and 2) redundancy, since there are typically many (1000) copies made. Some disadvantages are: 1) it is difficult to update just a section of the CD-ROM archive, 2) users must have a CD-ROM reader, and 3) while the data can be distributed the "expert advice" often needed to effectively use the data is difficult to distribute on the CD-ROM. Network access has advantages of: 1) it is easy to update selected portions of the archive, and 2) there are easy-to-use graphical user interfaces (GUI) providing access to many network archives. Some disadvantages are: 1) network connections can be slow or flaky, 2) the archive server can create single-point failure problems.

Hybrid CD-ROM/Network systems may offer the best of both worlds by providing distributed software, distributed data, and mature GUIs. Two examples of such systems are

NASA's Astrophysics Data System (ADS) and the Hubble Guide Star Catalog/IRAF GASP. ADS provides a mature GUI and allows for multiple distributed data servers, including local CD-ROM data servers. The Hubble GSC is distributed on CD-ROM, and can be easily access via the distributed IRAF GASP package.

4. ATRONOMICAL CD-ROM INNOVATIONS

A variety of "innovations" has been developed by astronomers in producing CD-ROM data products and archives. The EINSTEIN Slew CD-ROMs were the first to use the new FITS BINTABLE format in order to distribute machine-independent event-list data. Each EINSTEIN disk also includes four "flavors" of the documentation (Unix, VMS, MAC, DOS) in order to treat the end-of-line characters correctly, and also includes image display software on the disk for a variety of systems. To maximize the data per disk, the JPL group (Planetary Data Systems) has produced disks with compressed images and image de-compression software. To facilitate rapid scanning of the images, a browse disk, containing highly blocked but non-compressed images was included. The International Halley Watch group needed to produced images in both FITS and PDS data formats, and therefore split the data into headers (FITS and PDS) and image arrays. By concatenating the appropriate header and data users can read in either format. The Hubble Guide Star Catalog CD-ROM and the GASP package represent the first distributed access/CD-ROM package. The EUVE group has set a new record for rapid data CD-ROM distribution by producing their first CD-ROM of archive data less than 1 year after launch.

CD-ROM archives are not always limited to those which comfortably fit on a small number of disks. The JPL group is producing ~100 separate disks for the PDS archive, the STScI Plate Scans are being distributed on ~100 separate disks, and the International Halley Watch Archive is on a 28 disk set. One should keep in mind that these archives, while large, are dwarfed by many commercial applications.

The list of currently planned (or discussed) CD-ROM titles is fairly impressive: the VLA All-sky Survey, ApJS, IRAF Software distribution, to name a few. A (perhaps) un-anticipated beneficiary of astronomy CD-ROM archives is the educational system. Many secondary schools now have CD-ROM readers attached to PCs. Many teachers have discovered the abundance of astronomical images available on CD-ROM and are using the data in the classroom.

5. FUTURE TRENDS

In looking toward future changes in CD-ROM technology, one somewhat facetiously hopes there will not be any! This is because one of the strengths of CD-ROM as an archive medium is that it is NOT changing, and is therefore a stable standard. That said, some changes (improvements) to CD-ROM technology are occurring. Double and Quad speed drives have recently become available, drives which read the data even faster are likely. Multisession CD-ROMs are becoming common, these allow multiple ISO9660 file systems on a single disk. This will allow users to write data to the disk in small pieces, as it becomes available, instead of saving up 660 Mbytes until writing a disk. The cost of CD-ROM "interoperability" has been a least common denominator system - directories can be no more than eight levels deep, logical file links are not allowed, and filenames may have only an eight character prefix, one ".", and a three character suffix. These limitations are overcome by the new Rock-Ridge (RRIP 1993) extensions to the ISO9660 standard. These allow full Unix-like filenames, links, and directory structures while maintaining (in parallel) the older ISO9660 conventions.

CD-ROM using blue lasers, or with two active surfaces have been discussed, but both of these involve changes to the audio standard and would be incompatible with current systems. Large (200+ disk) jukeboxes have just become available. Developments in the commercial/ consumer market seem to indicate that CD-ROM will be around for some time. For example, CD-ROM is very often the medium for "multi- media" applications, all branches of the Federal Government continue to put more of their records on CD-ROM, Kodak is marketing a CD/WO system for home photography ("Photo-CD"), and USA Today on the day of this talk (5/17/93) announced an agreement between IBM and Blockbuster Video to use D/WO in order to make custom Audio CDs for consumers at Blockbuster stores. These last two CD/WO applications promise to reduce the price of CD/WO media from it's current ~$30/disk to ~$10/disk, therefore dropping the per-byte costs by a factor of three (See Table 1).

In closing, we note that optical tape is a new technology that offers a quantum leap beyond current optical storage devices. This technology is still in its infancy, but promises to offer up to a Terabyte of storage on a single tape, 60 second access to any part of the tape, at a cost of $0.01/MB (Owen 1992).

REFERENCES

Epchtein, N. 1993 in Workshop on Databases for Galactic Structure, A. G. D. Philip, B. Hauck and A. R. Upgren, eds., L. Davis Press, Schenectady, p. 97

Humphreys, R. M. 1993 in Workshop on Databases for Galactic Structure, A. G. D. Philip, B. Hauck and A. R. Upgren, eds., L. Davis Press, Schenectady, p. 87

Martin, M. and Rengarajan, K. 1993 NASA Science Information Systems Newsletter, 28, 31.

Owen, D. 1992 CD-ROM Professional 5, 73

Richmond, M. 1993 in Workshop on Databases for Galactic Structure, A. G. D. Philip, B. Hauck and A. R. Upgren, eds., L. Davis Press, Schenectady, p. 207

RRIP 1993, Rock Ridge Interchange Protocol, draft available via request to: cdtec@dgdo.Eng.Sun.COM

Sturch, C. R., Laidler, V. G., Lasker, B. L. and Postman, M. 1993 in Workshop on Databases for Galactic Structure, A. G. D. Philip, B. Hauck and A. R. Upgren, eds., L. Davis Press, Schenectady, p. 201

DISCUSSION

RATNATUNGA: With respect to the on demand, fast low cost audio CD ROMs that are expected to be produced by IBM in stores: Will this technology be compatible to the ISSO 9660 standard and have sufficient media quality for long life computer CD ROMs?

GARCIA: Yes, most probably. The technology is the CD/WO currently available, which is fully compatible with ISO 9660. The media quality and lifetime (50+ years) should also be sufficient for computer and archiving applications, but just as with standard CD-ROM and DAT/Exabyte tape, the quality will vary among manufactures. Only the best "computer certified" media should be used. As CD/WO is based on a gold substrate, rather than aluminum as for CD-ROM, is should not be subject to the oxidation sometimes seen with CD-ROM and may therefore have even a longer lifetime.

CREZE: There are a number of data management issues which must be cared for about the use of CD ROMs in addition to the technical aspects. (1) Concerning the distribution, there is already a tendency that specialized CD ROMs go straight on the disks of people specialized in the field. Casual users would just not know that they exist, or the effort for them to get the CD ROM data might discourage people from trying. (2) Concerning the archiving. One should have in mind that any standard may be abandoned at any time for purely commercial reasons. So someone should care for the data to be transferred to new media, which means the institutions in charge. The lifetime of such institutions may be much more crucial than the hardware lifetime of the medium.

GARCIA: Very true. As to 1) CD-ROM archive must be well advertised, in order to ensure that the appropriate community, not just the experts, knows of their existence. They must also be easy to obtain and easy to use, just like an archive or database. Two things which aid this are an index of the contents of the CD-ROM set, and an easily understood layout, for example one based on RA/DEC. As to 2), it is particularly true of NASA missions. The lifetime of the organization (ie, the mission team) is getting shorter and shorter due to budgetary pressures, and an increasing number of new missions. This makes it critical that plans for archiving be developed early in the mission, and that the archive contains more than "just the bits", but as much of the mission expertise as possible.

PIER: How do you submit data to the mastering plant?

GARCIA: Mastering plants aim to please, and will accept a variety of formats and media, including UNIX tar, VMS backups, FITS tapes, raw ISO9660 images, DOS files, and 9-track tape, DAT, Exabyte, SCSI hard disks, etc. It is best to check with the plant of your choice to see what they can easily handle.

PIER: How many FTEs does it take to produce a CD ROM archive?

GARCIA: The time required to generate the CD ROMs is small, typically a few weeks interaction with a manufacture is sufficient. The real effort goes into generating the archive itself, and is largely independent of the media. In the Einstein case, it took ~5 FTE/year for ~2 years to generate the archive. The majority of this effort went into verifying the data for completeness and accuracy, and preparing the necessary documentation.

ROMAN: For the ADC CD ROM, 1 1/4 FTE for two years were required to prepare the material for mastering after the catalogues existed. The time was used to check the catalogs for errors (and correcting them), preparing FITs headers, and ensuring adequate documentation in LATEX and ASCII format.

GRIFFIN: Despite the obvious advantages of these technological developments as tools to aid research, the extrapolation of the "home-ownership" trend may lead to the disquieting situation in which each astronomer becomes a self-contained mini-CDS and lacks the essential ability or feedback regarding updates, error corrections, etc. The situation prevailing up till today will then be replicated, though in digital instead of a printed medium.

GARCIA: There is clearly a continuing need for Data Centers, CDS included, to act at "centers of expertise" for the data sets in their care, and to issue health warnings, updates, and corrections for data sets. The technological developments (ie, CDROM, CD/WO) allow moderately large data archives to be distributed to the users cheaply, and therefore Data Centers do not necessarily have to be the only Centers where data is stored. Astronomers who collect data sets must then take care to ensure they keep up to date, much as astronomers who collect software (ie, IRAF, XSpec, MIDAS, etc..) must. On the Einstein program, we found there were a significant (and vocal!) number of users who wanted to become "mini-data centers", but this may not be the case for every program.

LASKER: With regard to the several comments made regarding the lifetime of archiving technology (12-inch worms, CD ROMs, etc.), the conclusion seems inevitable that the library or data center of the future will contain a laboratory for operating obsolete archive technology; i.e., not all materials will be transcribed onto newer technologies and at least a small amount of the older material will need to be read.

GARCIA: One can only hope to minimize the need for transcription by choosing technologies that become obsolete slowly. In considering optical disk technology, indications are that CD ROM will become obsolete relatively slowly.

LASKER: When ordering a production quantity of CD ROMs, the quantity is usually estimated on the basis of long-term needs and anticipated future requests. How should one store the ones stockpiled for future needs so as not to degrade the 25 yr life expectancy?

GARCIA: Storage techniques appropriate for archival photographic plates are believed to be appropriate for CD ROMs as well. Degradation is due to oxidation - therefore one would expect that storage in a dry nitrogen bath would be very effective. CD ROMs are expected to last for ~ 25 years or more under normal office conditions.

PHILIP: Is there any trend towards a common way of reading CD ROM disks? I have a number of astronomical catalogs, the Microsoft Bookshelf and every disk requires a different piece of software to run.

GARCIA: This proliferation of UIF and access software seems likely to continue. It is not a phenomenon restricted to CD-ROMS, network based archives (ie, HEASARC, NED, HST/Starview, SIMBAD, GRO Archive, etc..) also use different UIF and software. ADS aims to provide a common UIF for all of these archives. On the positive side, astronomy CD-ROMs maybe converging to a common format: FITS files and ascii indexes, which would allow common access software (ie, FITS readers). The HST/GSC and Einstein CD-ROMs both have IRAF based access software.

URBAN: A previous question asked about storage of CD ROMs affecting the 25 year lifetime. I stated that in a previous slide Dr. Garcia showed (concerning the lifetime of CDs) environmental factors (140 Deg. F, 85% rel. humidity) were used in the calculation.

GARCIA: The slide showed the results of an accelerated lifetest, which used artificially harsh conditions (140 Deg F, 85% humidity) to simulate many years of storage in normal office conditions. Predicted lifetimes varied among manufactures form 25 to 100 years.

BANIA: Hasn't AT&T just gotten an enormous increase in bandwidth? Hasn't this changed everything?

GARCIA: Perhaps, and it would be wonderful if it did, but I remain somewhat skeptical. There have been enormous increases in network bandwidth in the last few years, and they have been accompanied by matching increases in usage and data volume. The result from my personal point of view, at the workstation in my office, has been only minimal increase in long distance data flow. Recall that for Einstein, CDROMs provided 1000 times higher data distribution rate than the network.

CASERTANO: Regarding Dr. Philip's comment about ease of use of catalogs: as a user, I cannot say enough in favor of distributing the data in plain, old ASCII. Even the small changes between operating systems are easy to negotiate, and anybody can manipulate ASCII data. This is much more sensible than having the data on a fancy format, for which a new piece of software must be learnt - not to mention the possibility that the software may not be provided for the platform (DOS, MAC, UNIX) one is using. Please, leave the data in ASCII!

gravitational field and upon whether the system is in dynamical equilibrium. It is commonly considered that A and early F stars are not in dynamical equilibrium. Thus, the survey, although including A and F stars, is extended to late F and early G stars. The difficulties of selecting tracer stars for the K(z) study were outlined by King (1989).

2. THE DATABASE

This catalog (see Paper I) consists of four types of databases which include positions and pixel-magnitudes of 3155 A, F and G stars at the South Galactic Pole, MK spectral types, radial velocities and Strömgren 4-color and Hβ photometry. There are cross references to HD, SAO, Bok and Basinski number (1964) and 2DF spectral and CCD image numbers. This paper deals mostly with the spectral and photometric calibrations and errors of their determinations.

2.1 Positions and Pixel Magnitudes

A direct film copy covering approximately 40 sq-deg provided by the UK Schmidt Telescope Unit (UKSTU) at the Royal Observatory, Edinburgh, Scotland, was scanned using the Yale PDS 2020s microdensitometer by John Lee with a 40 x 40 micron aperture which yields a positional accuracy of approximately 2.6"/pixel. All positions were derived from the pixel X and Y coordinates with an AED model 767 image display using the BAFL program developed at Yale. The overall catalogue position accuracy is about ± 1.5" which is constrained by the half size of a pixel. Finding charts were produced by an Apple Laser-Writer for observation and identification purposes.

This 40 sq-deg. area was divided into 6 zones (named as SGP1 to SGP6), each containing 18 frames due to the size of the AED display dimension of 750 x 1000 pixels. Therefore, the 18 frames per SGP zone were delineated as SAxxxyy; the x's are pixel numbers in right ascension and the y's are pixel numbers in declination.

The magnitudes were derived from the PDS scan of a direct film IIIa-J emulsion with a gg-395 filter. The accuracy is about ± 0.2 magnitude (the error can increase to as high as ± 0.5 magnitude for images located near the edge of the film). The luminosity function for a sample of 2669 stars using this pixel magnitude peaks at 14.65 magnitude (Lu, 1991a).

2.2 2D-Frutti Spectral Classification

Nine hundred and sixty-one spectra have been obtained for 870 stars with the 2D-Frutti spectrograph at CTIO. The system is a combined image-tube and CCD system with 132 pixels in y and 3080 pixels in x and with a dispersion of 45 Å/mm at second order. Spectral and luminosity standards for A, F and G stars between various luminosity classes were obtained using the identical 2DF settings for the calibration of classification (Fig. 1). The accuracy in classifying in spectral type largely follows the signal-to-noise (S/N) ratio in the 2DF spectra. The mean difference is 0.81 ± 1.93 subclass for 71 HD stars (in Paper I) in all luminosity classes compared to the Michigan Catalogue of Spectral Types for HD Stars (Houk 1982). The mean spectral distribution for dwarf stars peaks at F9.3 V which agrees well with the photometric (b-y) color of 0.359 or F9 V.

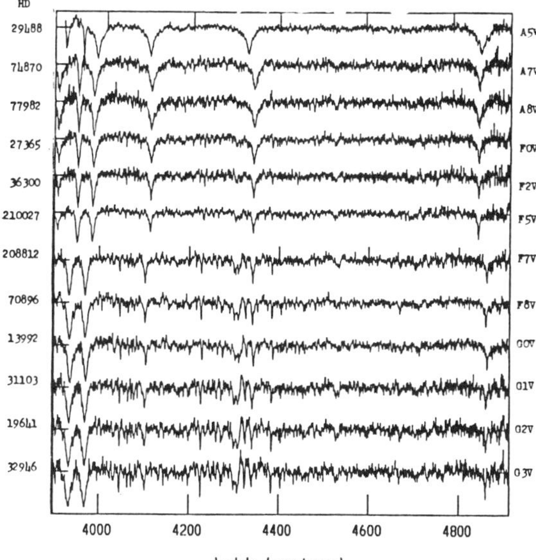

Fig. 1. NOAO/IRAF spectra for A5 V - G3 V stars.

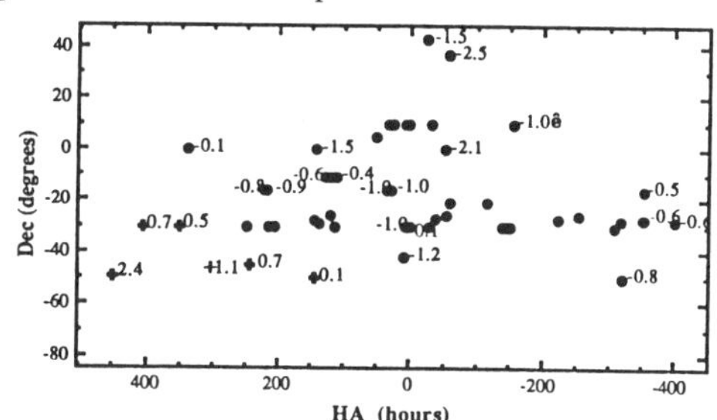

Fig. 2a. 2DF Spectrograph Pixel Shift in HA and Dec, 1989

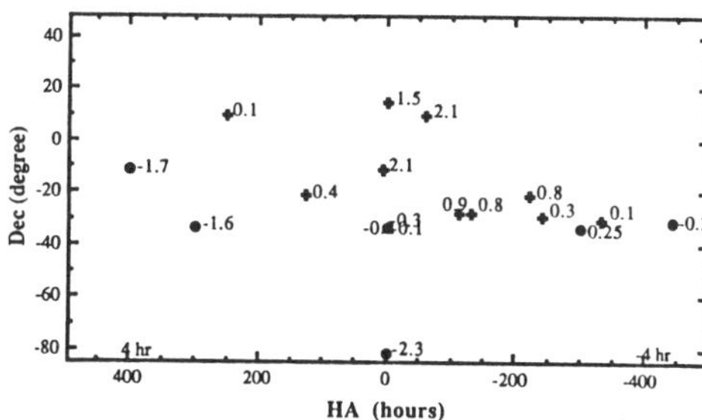

Fig. 2b. 2DF Spectrograph Pixel Shift in HA and Dec, 1990-91

2.3 Cross-Correlation Radial Velocity

Analyses to obtain radial velocities with an accuracy of about ± 5 km/sec at 0.5 Å/pixel using 2D-Frutti spectra are now in progress. Preliminary results were reported by Lu (1991a, b). Spectra with a very low signal-to-noise ratio will generally affect the accuracy and the determination of velocity using cross-correlation and the new technique reported by Carney et al. (1987) for extracting metallicities from high resolution and low S/N spectra. The test of fluctuation in the 2DF spctrograph using the comparison arc is shown in Fig. 2. The complete changeover between 1989 and 1990 was due to the changeover of the 2DF spectrograph. Radial velocity reductions using cross-correlation techniques were first done with the CCPROG program developed at Yale after considerable time and effort spent trying to use the early version of FXCOR in IRAF. The revised FXCOR seems to be working well in comparison with the RVSAO program except that RVSAO is more efficient and conducts the reduction non-interactively. The comparison between the revised FXCOR and RVSAO for three nights is shown in Fig. 3. The complete change over between 1989 and 1990 was due to the change-over of 2DF spectrograph. Of 685 stars, the weighted mean velocity is 13.7 ± 31.4 km/sec for 729 spectra (Lu 1991b).

2.4 Direct CCD and ASCAP Photometry

Distinguishing a subdwarf from a normal dwarf is essential in this study since there are about 7% metal-deficient or intermediate Pop II F stars that have been detected with the 2DF spectra. Photometrically, it is possible to identify them from their metallicity (m_1) and luminosity (c_1) indices using 4-color and Hβ photometry. Direct prime focus CCD images using the 0.9-m telescope with a TI chip and ASCAP (Automated Single Channel Aperture Photometry) measurements with the Yale 1-m telescope at CTIO were obtained for 1104 stars in the uvby, Hβ system. The observations show that it normally requires four to five minutes integration time to reach a 16th magnitude star in the vby bandpasses. However, the very low quantum efficiency at the u-band of the CCD chip required quadruple integrating time to achieve the same intensity. The ASCAP system is used for single object photometry, which is generally more suitable and efficient than CCD imaging using the Strömgren system since the u-bandpass in a photon counting system is about a factor of three faster than in the CCD system with comparable accuracy.

Neutral density filters of 2.5 to 7.5 magnitudes used for Crawford's bright standards (Crawford, 1975) caused some calibration problems because the transmission curves of the combined neutral density filter set is not well known. A systematic difference appears to exist in Twarog's secondary standards (Twarog 1984) and data from Olsen (1983) and McFadzean et al. (1982). A new set of HD stars fainter than 7th mag selected from McFadzean et al. (1982) were observed and calibrated as secondary standard stars within the 40 sq-deg of this catalog; residuals of these standards are shown in Fig. 4. The mean color distribution of (b-y) for 1104 stars peaks at 0.359 ± 0.080 which corresponds to a spectral type of F9 V. Spectral, color distribution and a typical (b-y) vs m_1 diagrams are shown in Paper I.

ACKNOWLEDGMENTS

We wish to thank Dr. Robert Williams and the staff at CTIO for all their support, and

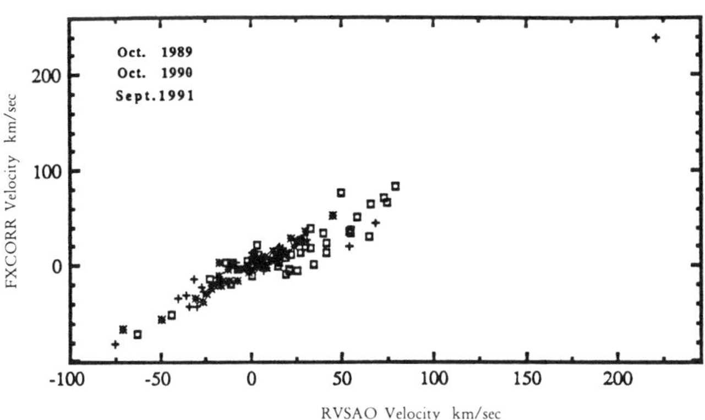

Fig. 3. RVSAO Radial Velocities vs FXCORR Radial Velocities

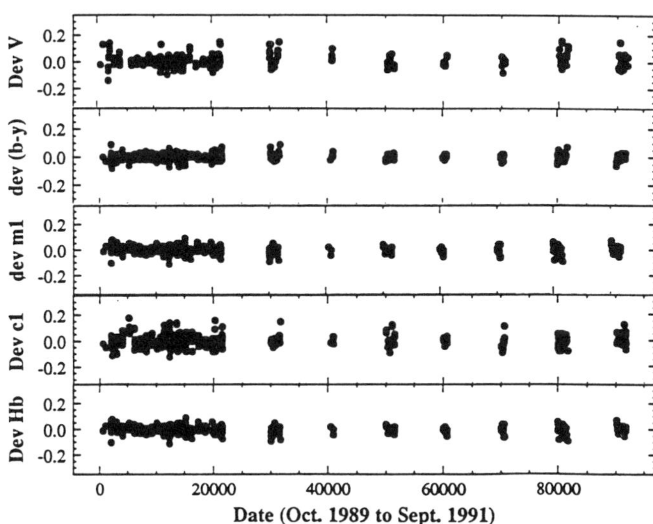

Fig. 4. Deviations of Standard Stars with Time.

the staff of UKSTU, Royal Observatory, Edinburgh, Scotland, who provided the needed film copies to conduct our earlier search for F stars. Special thanks are due to Dr. William van Altena for providing the PDS time and to John Lee who did all PDS scans at Yale, to Dr. Arthur Upgren, who has served as the project consultant, and as well as to Donald Platt, who has done all the radial velocity reductions. This research has been supported in part by grants from the National Science Foundation grant AST 8713183 and from the Connecticut State University system.

REFERENCES

Bahcall, J. N. 1984a ApJ 276, 156
Bahcall, J. N. 1984b ApJ 276, 169
Bahcall, J. N. 1984c in Astronomy with Schmidt-Type Telescopes, M. Capaccioli, ed., Reidel, Dordrecht, p 241
Bahcall, J. N. and Casertano, S. 1985 ApJL 29, L7
Bahcall, J. N., Hut, P and Tremaine, S. 1985 ApJ 290, 15
Bahcall, J. N., Flynn, C. and Gould, A. 1991 ApJ, submitted
Bok, B. J. and Basinski, J. 1964 Memoirs of Mt. Stromlo Obs. #16
Carney, B. W., Laird, J. B., Latham, D. W. and Kurucz, R. L. 1987 AJ 94, 1066
Crawford, D. L. 1975 AJ 80, 955
Houk, N. 1982 Michigan Catalogue of Two-Dimensional Spectral Types for the HD stars, Vol. 3, Univ. of Michigan, Ann Arbor
King, I. R. 1989 in The Gravitational Force Perpendicular to the Galactic Plane, A. G. D. Philip and P. Lu, eds., L. Davis Press, Schenectady, p. 147
Kuijken, K. and Gilmore, G. 1989a MNRAS 239, 571
Kuijken, K. and Gilmore, G. 1989b MNRAS 239, 605
Lu, P. K. 1988 in Calibration of Stellar Ages, A. G. D. Philip, ed., L. Davis Press, Schenectady, p. 217
Lu, P. K. 1989 in The Gravitational Force Perperdicular to the Galactic Plane, A. G. D. Philip and P. Lu eds., L. Davis Press, Schenectady, p. 95
Lu, P. K. 1990 in the ESO/CTIO Workshop on Bulges of Galaxies, Jarvis and D. Terndrup, eds., La Serena, p. 351
Lu, P. K. 1991a in Formation and Evolution of Star Clusters, K. A. Janes, ed., ASP Conf. ser. Vol. 13, p. 581
Lu, P. K. 1991b in Objective-Prism and Other Surveys, A. G. D. Philip and A. R. Upgren eds., L. Davis Press, Schenectady, p. 33
Lu, P. K., Miller, J. and Platt, D. 1992 ApJS 83, 203
McFadzean, A. D., Hilditch, R. W. and Hill, G. 1982 MNRAS 205, 525
Olsen, E. H. 1983 A&AS 54, 55
Oort, J. H. 1932 Bull. Astro. Insts. Neth. 6, 249
Oort, J. H. 1960 Bull. Astro. Insts. Neth. 15, 45
Philip, A. G. D. and Lu, P. K. 1989 The Gravitational Force Perpendicular to the Galactic Plane, L. Davis Press, Schenectady
Twarog, B. A. 1984 AJ 89, 523

DISCUSSION

PHILIP: Have you checked your spectral types for early-type stars against other star catalogs such as the Slettebak-Brundage Catalog?

LU: Yes, with the currently available photometry and 2DF spectra. About 3% of the 3155 catalog stars have spectral types F5 or earlier, (b-y) = 0.24 or bluer. There are 45 stars in common with the SB catalog and this study. The comparison of these hot stars will be discussed in a future meeting.

UPGREN: J. Stock and J. Rose have objective-prism radial velocities of about 10,000 stars at low galactic latitude and about 3000 stars at the SGP (error about ± 10 km/sec per star). They find many stars with $|RV| > 100$ km/sec in the first region and NONE at the pole. They lack any stars with $|W| > 100$ km/sec. It would be useful to compare your sample with theirs to see if the SGP has no high-W velocity stars, unlike at the NGP.

LU: I am familiar with the Stock and Rose project but do not have any of their data. I would be interested to look into their W velocities and compare them to our project.

RATNATUNGA: I wish to point out that PDS scans of 700 Å/mm objective-prism spectra have been successfully image processed to obtain spectral classification in the halo K-giant survey by Ratnatunga and Freeman in 1983.

ROMAN: I was interested in your spectral type distribution which shows a maximum for the giants near F8. Near the Sun, there are almost no F8 giants - it is in the Hertzsprung gap. Do you think this is due to metallicity? If not, how do you explain it?

LU: The maximum number of F8 giants is probably due to small number statistics and not a real distribution. The real peak distribution is probably at G2 or G3.

COHEN: Does anybody archive M giants near the SGP?

LU: Yes, Ken Freeman and his colleagues in Australia and Ken Yoss at the University of Illinois are conducting K and M giant studies.

WARREN: The scatter in your (B-V) vs. (b-y) diagram is rather unexpected. Could it be due to a mixture of stellar populations?

LU: Yes. The stellar populations in this study have not been separated.

JAHREISS: Is your database accessible to other astronomers?

LU: Yes. The database is available either by FTP or by Internet.

MERMILLIOD: Does your survey include the Blanco 1 open cluster (0^h, -30^m)?

LU: No, Blanco 1 is outside our 40 square degree region. It may be included in the future 100

THE DATABASE FOR GALACTIC OPEN CLUSTERS (BDA)

J. -C. Mermilliod

Institut d'Astronomie, Universite de Lausanne

ABSTRACT: A database, containing quite complete collections of observational data (photometry, spectroscopy, astrometry) and an extensive bibliography for stars in galactic open clusters has been developed since 1987 at the Institut d'Astronomie de l'Universite de Lausanne. It not only aims at storing data, but also at providing a working environment covering most aspects of the study of open clusters. The largest effort has been spent in solving the identification problems raised by the definition of so many different numbering systems.

1. INTRODUCTION

Decades of observations of stars in open clusters have produced a large amount of data differing in quality and degree of completeness. Consequently the analysis and study of open clusters is time consuming, because there are many tasks to perform: one has to collect and confront the data and determine a list of member stars before being able to analyze a cluster. Furthermore one is facing the serious problem of star identification, due to the proliferation of numbering systems: membership probabilities deduced from proper motions are never published with the same numbering system as the photometric data! Projects dealing with large number of clusters require the automation of as many processes as possible: manual work is clearly no longer feasible.

Since 1976, I devoted a lot of effort to compile observational data for stars in open clusters and took special care in preparing data files with uniform numbering systems for each cluster. The status of the data collections has been presented by Mermilliod (1985), but since that time the collections have been kept up-to-date, the work being now much easier due to the existence of the database.

2. CONCEPT

The design of the database has been tailored for the specific domain of star clusters. They are basically considered as collections of several tens to several thousands of stars located in limited regions of the sky. Working on clusters often consists in handling one kind of data for a sample of (member) stars, rather than many data for a single star. The database has been designed to be not only an efficient tool to store and retrieve data, including most functionalities it implies, but also to provide a versatile environment to analyze the data and develop the astrophysical study of open clusters.

3. STRUCTURE

It results from the above considerations that a cluster is the basic unit of the database. The database structure fully uses the directory hierarchy supported by the Unix system. Each cluster

Workshop on Databases for Galactic Structure
A. G. Davis Philip, Bernard Hauck and Arthur
R. Upgren, eds. p. 27 - 34 © 1993 L. Davis Press

J. -C. Mermilliod
Institute of Astronomy
University of Lausanne
Lausanne, Switzerland

forms an independent directory identified by its name and contains the available data in distinct files. The files are organized sequentially, compressed and sorted by star number and reference. Whenever possible, the records have the same structure: star identification, reference, data. A more detailed description has been published by Mermilliod (1988).

The basic concept of this realization is the attribution to each star of a unique number. The cross-identifications are systematically kept in tables (Mermilliod 1979) of which Table 1 shows an example. The nine references correspond to the nine columns (1 to 9). The first column contains the adopted numbering system, usually identical to that of the first reference. If data sets were to be included in the database in their original form, any structure or tool would be useless. Therefore, it appears fundamental to determine the cross-identifications of any new study with all the previous ones, to be able, for example, to recover positions, spectral types or membership probabilities.

TABLE 1

The beginning of the cross-identification table for NGC 2287

No	1	2	3	4	5	6	7	8	9
0001	1		4	7	58	103			70
0002	2	6	8	12	68	71	25	73	72
0003	3		16		66	77		72	
0004	4		25		65		24	74	
0005	5		15		69	79		77	
0006	6		18		67	88		80	
0007	7		93			94			
0008	8		82			80			
0009	9		62						
0010	10		58			93			

1) Cox A. N. 1954 ApJ 119, 188
2) Hoag A. A., et al. 1961 Publ. US Nav. Obs. 17, 347 (Pe)
3) Hoag A. A., et al. 1961 Publ. US Nav. Obs. 17, 347 (Pg)
4) Loden L. O. 1957 Ark. Astr. 2, 39
5) Ianna P. A., Adler D. S. and Faudree, E. F. 1987 AJ 92, 347
6) Andersen T. B. and Reiz A. 1983 A&AS 53, 181
7) Geyer E. H. and Nelles B. 1985 A&AS 62, 301
8) Gould B. A. 1897 Cordoba Photographs, cluster IV
9) Amieux G. 1988 A&AS 76, 305

4. CONTENT

All data collected so far have been installed in the database. New data found in the journals or received by email are regularly introduced in the database. Table 2 gives an insight of the database's present content (May 1993): it lists for each kind of data the number of clusters involved, the number of measurements and the number of stars concerned, except for the references and bibliography where the number of entries is indicated. The completeness of the database should be rather high for most kinds of data, and is still improving. Mermilliod (1992a) gave a report on the recent introduction of new data in the database. All data are referenced.

5. BIBLIOGRAPHY

The database offers also an extended bibliographic service. The Catalogue of Star Clusters and Associations (Alter et al. 1970) and its first Supplement (Ruprecht et al. 1981) have been incorporated to the database (one file per cluster). The 1969 - 1992 bibliography based on chapter 153 of Astronomy and Astrophysics Abstracts has been entered and can be interrogated with up to three keywords simultaneously. The information from the Fifth Catalogue of Open Cluster Data (Lynga 1987) has also been included in the database.

6. SOFTWARE

The file organization and their small size, a few hundred records, makes the consultation very rapid with Unix tools. The shell is also used to build menus which are usually arranged according to the sequence of steps corresponding to the work to be performed.

C codes are used for the development of "system" functions relying on C libraries, for example the up-date or graphic facilities, while most application codes adapted from existing programs available from various sources are written in Fortran.

The capability of comparing data from various sources is fundamental. This is especially important for the UBV photometric data which are by far the most numerous (see Table 2). The database offers extensive facilities to check the quality of UBV data. The most important software in the database is the one developed to plot and analyze the photometric diagrams in the UBV and Geneva systems. It offers many functions, and, among them, the selection of members based on the proper motion probabilities, or on the distance from the cluster center, the fitting of theoretical isochrones (computed on line), the photometric deconvolution of binaries. A more detailed description of these facilities has been published by Mermilliod (1992b).

Two calibration codes have been implemented on the database: one for the DDO system (valid for the red giants) and one for the uvby system. Improvement and extension to other photometric systems are planned.

7. CLUSTER MEMBERSHIP AND EXPERT SYSTEM

The membership estimation is one of the primary question to be solved. This action usually requires a complex interrogation of the database to collect all the available data and the examination of photometric diagrams to evaluate the position of the stars with respect to the main sequence or red giant region. Again, due to the number of clusters and number of stars in each cluster, this becomes too long to be done star by star. An expert system, adapted

TABLE 2

Database Content

Subject	Clusters	Measures	Stars
Identifications:	361		9752
Transit Tables:	179		63606
Coordinates:	419	28903	25103
Positions (round off):	425		37108
Positions (x,y):	373	128177	118022
Double stars:	192	1657	1225
UBV photoelectric:	409	29853	20271
UBV photographic:	263	88581	69498
UBV CCD:	57	26147	24762
RGU (pg):	73	9911	9911
Geneva 7-color:	184		4306
uvby measures:	154	5265	3732
uvby mean:	157		3423
uvby Eggen:	42	876	775
uvby CCD:	3	276	276
Hβ measures:	213	5463	3629
Hβ mean:	220	3312	
DDO:	125	936	755
Washington:	59	519	493
Walraven:	58	1432	1382
RI (Johnson):	2	405	350
RI (Kron):	4	1254	1111
RI (Eggen):	14	118	118
RI (Cousins):	15	487	487
RI (Cousins) CCD:	5	2470	2470
JHK:	32	729	727
Vilnius:	29	519	495
MK types:	257	7457	4495
MK types selected:	162		3974
HD types:	290		8351
Vsini:	76	2177	1565
Vr mean:	61	1777	1590
Vr individual:	181	29058	3406
Vr GPO:	9		505
Vr RFS:	7		141
Orbits:	35	197	175
Probability (μ):	57		21245
Probability (Vr):	2		87
Remarks:	233	3626	3085
gK stars:	229		3225
Am stars:	36		109
NM:	94		2645
Biblio Alter et al.:	587		20000
References:			2500
Bibliography (69-92):			2800

from a commercial software named Jonathan (Frot 1988), has been interfaced with the database, in which it finds the data facts. A set of 160 rules are used to infer the conclusion on the star membership. The possibility of getting a justification of the deduction made is one of the interesting features of this system.

8. PROJECTS AND USEFULNESS OF THE DATABASE

The database has been mostly used up to now to produce color-magnitude diagrams with isochrones to check new evolutionary models (Meynet et al. 1993). The preparation of a new atlas of color-magnitude diagrams for all observed clusters remains a long term project, but it may require a wider collaboration due to the increasing number of new observations to be discussed and analyzed.

The database may also be very useful to select clusters offering the largest numbers of data or, on the contrary, to prepare new observation campaign. For example, 139 NGC and 9 IC clusters do not have any UBV photoelectric data, as well as 85 of the 240 anonymous clusters installed in the database.

9. DISTRIBUTION OF THE DATABASE

The database is installed on a Sun Sparc 2 workstation and needs about 30 Mb of disk space. It contains about 6000 files. A copy is maintained at Meudon Observatory (France) on a DEC station thanks to the help of Dr F. Arenou. Due to its conception, it is not portable to non Unix systems.

The database is now sufficiently developed so that it may be distributed to interested colleagues working on open clusters. The entire database can be obtained on Sun or DEC cartridges. Data for any cluster can also be sent by email (mermio@scsun.unige.ch), together with references and bibliography.

10. DISCUSSION AND CONCLUSION

This system has been conceived for an application specific to open clusters and is based on the facilities offered by the Unix system. It is clear that the same structure can be used to store data for globular clusters or clusters in the Magellanic Clouds. But I will not extend the database content to these kinds of clusters, due to the amount of work already needed to maintain the database. However it could be a starting point for new extensions in these directions.

I hope that some kind of collaboration will be developed around the database to improve its content and enlarge the number of applications connected to it. In this respect I am very glad to thank all colleagues who have answered to my request to make their data available in computer-readable form and sent by email.

REFERENCES

Alter G., Ruprecht J. and Vanysek V. 1970 Catalogue of Star Clusters and Associations,

    Akademiai Kiado, Budapest
Frot, P. 1988 3 Systemes Experts en Turbo C, Sybex, Paris
Lynga, G. 1987 Centre de Donnees de Strasbourg
Mermilliod, J. -C 1979 A&AS 36, 163
Mermilliod, J. -C 1985 Mem. Soc. Astron. Italiana 56, 549
Mermilliod, J. -C 1988 Bull. Inform. CDS 35, 77
Mermilliod, J. -C 1992a Bull. Inform. CDS 40, 117
Mermilliod, J. -C 1992a in Astronomy from Large Database II, A. Heck and F. Murtagh,
    eds., ESO Conf. and Work. Proc. no 43, 373
Meynet, G., Mermilliod, J. -C. and Maeder, A. 1993 A&AS 98, 477
Ruprecht, J., Balasz, B. and White, R. E. 1981 Catalogue of Star Clusters and
    Associations, Supplement, I. B. Balasz, ed., Akademiai Kiado, Budapest

DISCUSSION

HUMPHREYS: That is a lot of work. How does one access this information?

MERMILLIOD: Presently there are two possibilities. (a) Get a copy of the whole database. It needs about 30 MB of disk space and is better suited for a workstation. (b) Get by email the data you are interested in. I have prepared a facility to extract selected data, with references, from the database and send the resulting file by email. Both facilities are used. Some 15 copies have been distributed and I provided several colleagues with the data they needed.

PLATAIS: We have a copy of the BDA at Yale University Observatory. The database is extremely useful for different studies in open clusters and is convenient in use. I'm impressed by the mammoth amount of job put into the BDA development. The BDA could be recommended to anyone working with open clusters. Since several copies of the BDA have been released elsewhere around the world I would like to ask how to keep all these copies equally updated? How do your maintain feedback between your master version and each particular copy of BDA?

MERMILLIOD: I am keeping a journal, namely a list of up-dated new files with the date of the modifications, so that it is possible to know all the changes from a given date. I am maintaining a copy on a DEC station in Paris and for the last update I sent by FTP a compressed tar file with about 760 individual files. So in principle, there are no real problems to up-date remote copies.

EICHHORN: I have a suggestion to make the database available on the Astrophysics Data System. This will allow one to always have the latest version on-line and directly available to the researchers. There may be the possibility for an ADS mode to put the database on-line and receive regular updates from the authors.

MERMILLIOD: Yes, this would be an interesting possibility. But you probably will lose most of the facilities offered by the database, especially the graphic ones.

GRIFFIN: Where is the financial support for this extremely valuable public service? One thinks too of the AAVSO, whose effort is magnificent but may ultimately be limited by the private resources it can attract. Should the financial support for all these commendable efforts not become, at least in part, a greater concern for the astronomical community at large?

MERMILLIOD: The development of this database is motivated by my interest in open clusters and the research I am doing. Therefore it is part of the scientific program for which we have a grant from the Swiss National Funds (FNRSI). More than financial support, I would appreciate a deeper collaboration to develop the database, i.e. receiving more data by email. It would be also very important if more people would accept to share their application software.

HAUCK: Approximately one-third of the funding, not only for the star cluster database, but for our general catalog of photometric data and other research, is provided by the Swiss National Fund for Scientific Research. The other part is provided by the Lausanne University.

CREZE: You showed a facility that splits a binary star into two components using the joint photometry and the cluster isochrone. Is there always an exact match and is the solution unique?

MERMILLIOD: There are two conditions to be filled: (a) both components should be close to the isochrone and (b) the binary loop computed from the primary by adding "ideal" stars from the isochrones should cross the observed binary postion. So the solution should be quite reliable.

PHILIP: The photometric systems used in the database - is it the classical distribution with UBV studies the most numerous and Strömgren and Geneva photometry in the next place? What is the distribution of measures by photometric systems?

MERMILLIOD: The largest amount of data have been obtained in the UBV system (See Table 2). Next, both the uvby and Geneva seven-color system share the second position. Other systems are far behind.

RATNATUNGA: I noticed that proper motions were not mentioned in most of your talk. Are they not included in your database?

MERMILLIOD: Up to now, only membership probabilities based on proper motions (57 clusters) have been included in the database, because, initially I did not intend to redo a detailed analysis of proper motions. But if there is a need for inclusion of the proper motion data there is no technical difficulty.

RICHMOND: Regarding proper motions: do you keep separate (x,y) positions measured by different observers for stars in the same clusters?

MERMILLIOD: No. x, y positions are not meant for astrometric purposes but for making the best census of stars in the cluster field according to the studies made. So we keep only a single position per star.

ROESER: In your data base you have a large number of stars, but a low number of identifications. What, exactly, do you understand by "identification"?

MERMILLIOD: Under identification I mean cross-identifications with large astronomical catalogs like HD, DM, BS, GCVS, IDS, NLS and LSS.

OMONT: Could you comment more about the degree of completeness of the knowledge of open clusters in the Galaxy, and also of what can be expected from the development of infrared studies of open clusters?

MERMILLIOD: About 1200 open clusters have been identified from all-sky surveys (POSS, ESO). Less than 50% have yet been investigated photometrically. The situation for remote, faint clusters is improving thanks to CCD observations. Infrared appears very important for low-mass member detection and study and for embedded clusters which are generally not detected in visible light.

THE NEW WASHINGTON VISUAL DOUBLE STAR CATALOG (1994)

Charles E. Worley and Geoffrey G. Douglass

U. S. Naval Observatory

ABSTRACT: Following the 1984 version of the Washington Visual Double Star Catalog, we began planning a successor to be issued approximately ten years later. The 1984 catalog was intended primarily to provide an updated and corrected version of the Lick Index Catalogue, and was produced in the same format, though with considerably augmented notes and a bibliography. It now appears that we will be able to meet our tentative schedule, and this report details both the changes in format and in content, as well as some of the problems we have encountered in melding other data sources into our catalog.

1. INTRODUCTION

This discussion represents in part an abridged, but necessary repeat of information presented by one of us recently at a conference held at Pine Mountain, Georgia (Worley, 1992). Differential measures of the components of visual binary stars of sufficient accuracy, still to be useful by modern standards, began about 1820, and continue today. The number of known pairs has grown from a handful at that time to more than 77000 now. Astronomers engaged in this work have always been cognizant of the need not only to preserve the data, but also to disseminate it to the wider astronomical community. Initially, this response was accomplished spasmodically by the publication of catalogs, and this necessarily resulted in a product which was obsolescent (and often badly so) when received by the user.

Thus, Aitken's famous New General Catalogue (Aitken 1932) had a closing date of 1927.0, but was not published until late in 1932, which meant that for many users the data was already more than six years out-of-date. To a lesser degree, the same remark can be made about the Lick Index Catalogue (Jeffers and van den Bos, 1963), and the Burnham Catalogue (Burnham, 1906), although in the latter case a rather inelegant out-of-order appendix served to keep the data more current.

In preparation for the issuance of the Lick Catalogue, the data contained on their handwritten cards was transferred to punch-cards beginning in 1955. At about the same time the data south of -30 degrees declination, long collected by van den Bos, was added to the Lick material, thus forming a complete all-sky double star catalogue for the first time.

2. AT THE NAVAL OBSERVATORY

In 1965 the Lick cards were transferred to tape, and the Naval Observatory assumed responsibility for the future of this project. We have made many corrections of all kinds, and kept the observational data base up-to-date. We have also collected (at this point) most of the earlier, pre-1927, measures not included by the Lick effort, so that we are now approaching a

Workshop on Databases for Galactic Structure
A. G. Davis Philip, Bernard Hauck and Arthur
R. Upgren, eds. p. 35 - 38 © 1993 L. Davis Press

Charles E. Worley
U. S. Naval Observatory
3450 Masachusestts Ave.
Washington, D. C.
20392

complete file. We also gradually have developed powerful software, both for our own uses and to be able to provide other users with data in a timely fashion. Hundreds of such requests have been filled.

We produced a new tape of the IDS in 1976. This was a semi-updated version incorporating corrections and new pairs only. Storage of data on punch-cards is awkward and time-consuming to change, and we felt it too daunting a task to do more with our limited resources. Fortunately, the situation was radically improved in 1982, with the advent of terminals connected directly to the computer and its disk storage. One of us then began the updating of certain portions of the data, concentrating on motions and additional measures. Few alterations were made to the magnitudes, spectral types, and proper motions. The Notes were rerevised and considerably augmented, and a bibliographic index provided. Of course, all known corrections were included. The format itself, however, remained virtually identical to the Lick version. The WDS is now available from the data centers on tape, and also from the Astronomical Data Center (1991) as a CD-ROM.

3. TOWARDS A NEW CATALOG

Work began in 1984 on a comprehensively revised and reformatted catalog. The principal changes are enumerated below:

1. A single position is given, and is intended to represent the epoch J2000 whenever possible, rounded to the nearest 0.1m in Right Ascension, and nearest minute in Declination. Proper motion corrections are employed when significant. One should note, however, that positions for many pairs are only crudely known.

2. Numbered discoverer designations are being provided for all objects, because they are the only common denominator available for all pairs. Other designators, such as durchmusterung number, exist for only a fraction of the pairs, while the use of positional designations is likely to result in confusions due to the aforementioned inaccuracies, as well as eventual crowding as the data base increases.

3. Component designations have been revised.

4. Information regarding dates of observation, initial and last position angles and separations, and total number of measures, was revised for the 1984 edition, and has been kept current.

5. Wherever possible, Johnson V magnitudes are quoted. If, in our judgement, the magnitude appears to be a combined value, then we have first consulted an unpublished catalog of visual magnitude differences; otherwise, we have used values estimated by reliable observers.

6. Two-dimensional MK spectral types are used. These are both from the Michigan Spectral Survey and earlier sources. For other pairs we use HD or whatever other types are available.

7. Proper motions (and for that matter) positional data have been taken from the ACRS (Corbin and Urban, 1991) when possible. Otherwise, they come from a variety of sources.

8. Revisions to the Notes have continued, aimed at making them more astrophysically relevant. Some of the less important information may well be relegated to specific subsidiary files, though no final decision on this has yet been made.

9. Corrections and additions to the Bibliography continue to be made, and a current version will be incorporated in the final product.

4. DATA PROBLEMS

Much of the recent updating of the WDS has been by means of the invaluable SIMBAD files of the Strasbourg Observatory. During use of this material, however, a surprising amount of missing data concerning binary stars was encountered. As an example, we have arbitrarily examined the files for R.A. eleven hours. The results are tabulated below.

TABLE 1

WDS (eleven hours)

Total Pairs	Durchmusterung Pairs	Pairs Missing in SIMBAD	Pairs Unidentified in SIMBAD
2129	1437	256 18%	238 17%

While it is somewhat distressing to find so many pairs simply missing from SIMBAD, the real danger for the user are the many included but not flagged as being binary. Consequently, we are carefully noting such cases (as well as other problems found) and will in due course transmit these data to Strasbourg. Similarly, in adding the Michigan Spectral Survey types, we have found many unflagged pairs. Therefore, we have prepared and sent to its authors a cross-index of WDS versus Durchmusterung number. In the case of the PPM catalog, the authors were mislead into use of the obsolete 1976 IDS tape. As a result, there are many unidentified binaries, including some ranked "high precision".

The lesson from these three examples is clear: one needs to exercise due care when dealing with astronomical data bases involving substantial numbers of binaries. Otherwise, both the compilers and the users are apt to receive some unpleasant surprises.

5. THE FUTURE

Barring unforseen circumstances, we are confident that we will be able to issue the new version of WDS next year (1994). It will be distributed by the data centers in whatever electronic form is deemed appropriate by them. Hopefully, this will also include a CD-ROM. A related project will entail the construction of a new catalog of visual binary orbits thereafter.

A longer term problem is the eventual care and disposition of the WDS database. We have, at present, no solution to this quandary.

REFERENCES

Aitken, R. G. 1932 New General Catalogue of Double Stars within 120 Degrees of the North Pole, Carnegie Institution of Washington
Astronomical Data Center, Goddard Space Flight Center. Selected Astronomical Catalogs Vol 1, 1991
Burnham, S. W. 1906 A General Catalogue of Double Stars within 121 Degrees of the North Pole, Carnegie Institution of Washington
Corbin, T. E. and Urban, S. E. 1991 Astrographic Catalogue Reference Stars, Washington, U. S. Naval Observatory
Jeffers, H. M., van den Bos, W. H. and Greeby, F. M. 1963 Index Catalogue of Visual Double Stars, 1961.0. Pub. Lick Obs., XXI
Worley, C. E. 1992 in Complementary Approaches to Double and Multiple Star Research, IAU Colloquium 135, H. A. McAlister and W. I. Hartkopf, eds., ASP Conference Series 32

DISCUSSION

JAHREISS: The statistics you presented - was it done with the 1984-version or with the updated present version?

WORLEY: Updated.

JAHREISS: So, it isn't astonishing that the doubles are missing (in SIMBAD or PPM)?

WORLEY: Actually, most of the existing discrepancies were already present in the 1984 version.

MERMILLIOD: Where did you take the UBV data from? SIMBAD UBV data are complete until the end of 1986. We have new UBV data to the end of 1992 that could be useful to you.

WORLEY: The UBV data were taken from SIMBAD or other sources. I will be glad to have the new data!

CREZE: I am not at all surprised by the incompleteness of double star information in SIMBAD. This is an issue we have to work on in the near future, in cooperation with you.

WORLEY: I agree. Some time ago I delivered a copy of the first six hours R. A. discrepancies to Strasbourg. Another 8 hours are ready.

COHEN: Returning to the issue of SIMBAD's incompleteness, we have recently discovered that SIMBAD does not include Buscombe's annual supplements of UBV and MK spectral types.

ADC AND OTHER NSSDC DATA SOURCES

Nancy Roman

NASA/Goddard Space Flight Center

The U.S. National Space Science Data Center (NSSDC) contains extensive archives of astronomical satellite data of potential interest in galactic astronomy. Most of the ground-based catalog data relevant to this field resides in the Astronomical Data Center (ADC). This is one of a world-wide affiliation of data centers of which the first was the Centre de Donnees Astronomique de Strasbourg (CDS) in France. The one at NASA, Goddard Space Flight Center is almost as old. We exchange data regularly with the CDS and, to a lesser extent, with the other centers. Currently we have an archive of about 675 catalogs, most from ground-based data. We obtain new catalogs both directly from the compilers and through our exchanges. We also create new catalogs by combining data from several sources and we modify catalogs both to correct obvious errors and to make them more easily useful. Recently, both the CDS and NASA have been archiving smaller data sets from publications which are large enough to make electronic access desirable but that are well below book length. The American Astronomical Society (AAS) is planning to put all such data sets from their publications (the ApJ, ApJS, ApJL and the AJ) on CDROMs. Hence, normally, we shall not make an effort to archive these tables but we will attempt to obtain moderate size tables from other North American publications as well as those from the AAS journals which are specifically requested.

The catalogs are divided into eight categories with numbers assigned by the CDS. The categories, the number of holdings as of mid April, 1993, and examples of the types of catalogs of particular interest in galactic structure problems are as follows:

Astrometric - 134 - Parallax, Yale Zone, ACRS, PPM, Luyten, and the durchmusterungs

Photometric - 139 - Individual stars on various color systems including the UV, IR and 100 microns, and polarization; cluster stars, stars in localized regions, special types of stars, and interstellar band data

Spectroscopic - 131 - HD, HDE, Michigan; radial and rotational velocities; surveys of various regions and particular types of stars; and data on interstellar bands and lines and on interstellar extinction

Cross Identifications - 35 - Durchmusterungs, double stars, SAO, BSC, AGK3, variable stars, IR sources

Combined and Derived Data - 65 - Space velocities and locations for nearby stars, high velocity stars, various types of stars, stars in various regions such as clusters, associations, and the halo, maser emission, and X-ray objects

Workshop on Databases for Galactic Structure
A. G. Davis Philip, Bernard Hauck and Arthur
R. Upgren, eds. p. 39 - 42 © 1993 L. Davis Press

Nancy G. Roman
NASA/Goddard Spaceflight Center
Greenbelt, Maryland
20771

Miscellaneous - 30 - Photometric sequences, double stars, bibliographies, finding lists, indices, constellation boundaries

Non Stellar - 113 - Open and globular clusters, molecular clouds, associations; nebulae including dark, emission, reflection, planetary, supernovae remnants, H II regions, and globules; CO distribution, IR data, gamma-ray data, pulsars, NGC and the revised NGC

Radio - 17 - 3CR, 4C, sky surveys in H I, L-band and centimeters

The radio list is short at present but a major effort is being undertaken to expand that list.

In addition to the data available electronically, 114 of the most requested catalogs are available on a CDROM.

The easiest way to access these data is to login to NSSDCA.GSFC.NASA.GOV with the user name NODIS (NASA On-line Data & Information System). No password is required. Clear directions lead the user through the process of locating data of interest and of acquiring that data electronically. NODIS contains various options (some of which are discussed below). One is the ADC On-line Information System. After requesting this option, the user is asked how the data search should be conducted. Often a search by key word or short title (which may be the author's name) is the easiest. To obtain a complete list of the catalogs of a given type (e.g. astrometry or radio astronomy), use the search by ADC number. For many catalogs, a brief description of the contents is available in addition to the title. The data can be requested by following directions. They will be staged for pickup and usually are ready in a few minutes. A message will be sent telling the requester that the data are available and where they are located. The catalog can then be retrieved by FTP. Data can also be acquired on magnetic tape or, for short catalogs, in hard copy form by contacting:

(for US requesters)

NSSDC Coordinated Request and User Support Office (CRUSO)

(or, for other requesters)

World Data Center-A for Rockets and Satellites. The address for both sets of requesters is Code 633, NASA/Goddard Space Flight Center, Greenbelt, Maryland, 20771, U.S.A.

Several of the other NODIS options may also be of interest to astronomers. The Master Directory provides information about data not held by the NSSDC as well as those that are. Another category provides access to other on-line data. It should be noted that these are more limited than it might appear. The HST data are only the early release data; the remainder must be requested from the Space Telescope Science Institute. The IUE data must be accessed by exposure number. The NRAO data are only the 1400 MHz and 4.85 GHz maps. The HEAO2 data are from the IPC and the HRI. The HEAO3 data are from the anti-coincidence shield. The IMP data are for cosmic rays, and electric and magnetic fields. The Master Directory can be consulted for the much more extensive holdings.

In addition to the ADC CDROM, several other CDROMS of data of interest to astronomers are available from the NSSDC. Of these, data from the Einstein Observatory and the IRAS Sky Survey Atlas are most likely to be of interest in galactic structure research. A catalog of the complete list of available CDROMs is available from the above addresses.

Within the NASA Space Science Data Operations Office, the Astrophysics Data Facility, a sister organization to the NSSDC, is preparing an on-line collection of abstracts of articles from the major US astronomical journals, the Monthly Notices of the Royal Astronomical Society and the Journal of Geophysical Research and its supplement. Eventually, it is expected to make page images available. The effort, called STELAR (Study of Electronic Literature for Astronomical Research), is a joint undertaking of NASA, the AAS, the ASP, the journal publishers, and a number of research libraries. Currently the abstracts cover the period from 1965 to nearly the present. The system also provides electronically submitted abstracts for AAS meetings, the AAS job register, and access to the NSSDC's near-line archive data (via NDADS - ARMS).

Although most ADC holdings are available only in machine-readable form, some of the older holdings are available on microfiche. Recently, we produced eleven hard copy volumes: the three durchmusterung catalogs and a cross index between the Washington Double Star Catalog, the Aitken Double Star Catalog numbers, the durchmusterungs, and the HD numbers. Although many astronomers now prefer to work solely with electronic catalogs, there appeared to be sufficient interest to make this printing desirable.

The ADC distributes a quarterly Electronic Newsletter describing changes to our holdings and other news. To subscribe, send Email to LISTSERV@HYPATIA.GSFC.NASA. GOV and put the command: SUBSCRIBE ADCNEWS Full name (the subscriber's real name) in the body of the message, not the subject line.

DISCUSSION

CREZE: What is the current delay between publication date and abstract availability in STELAR?

ROMAN: I am not sure. I think it is several months.

RICHMOND: Is that several months before the abstracts appear in ApJ, or after?

ROMAN: After - the abstract system doesn't quite keep up with the publications.

GARCIA: How are the "STELAR" abstracts related to the "RECON" abstracts?

ROMAN: They are the RECON abstracts.

PHILIP: Nancy, at a previous meeting, we discussed your project of publishing copies of the Durchmusterung. I gather that this project is now complete?

ROMAN: Yes. There is a prepublication copy of one volume on display. I was told they would be out early in May but they had not arrived by last week.

A PROPOSAL FOR A CENTER FOR SPECTROSCOPIC STUDIES

R. Elizabeth Griffin

Institute for Astronomy, Cambridge (UK)

Stars are key building blocks of galactic systems, and spectroscopy is a backbone subject, fundamental to astrophysics. Good-quality results require good-quality observations; spectrographs with a high accuracy capability have been available for three-quarters of this century, and continuing efforts are being made to explore ways of increasing the detail and the fidelity of the recordings. However, recent changes in the balance of effort going into observational astronomy are having unfortunate repercussions on stellar research. The current nature of research funding is adding to the imbalance, since the requirement of academic researchers to publish *results* tends to deny the scientific community the opportunity for large-scale surveys and collaborations, and to leave any coordination of effort to the enterprise of individual volunteers.

Many basic problems could be enlightened through the provision of homogeneous, high-quality surveys of stellar spectra. For example, one remarkable failure of twentieth-century astronomy is the confusion over the nature of the two bright A-type dwarfs Vega and Sirius; the cause seems to be a general lack of knowledge about quiescent A-type dwarfs, and it is both surprising and uncomfortable, at this stage in the development of astronomical research and in a world rich in large telescopes, to be able to identify such a major deficiency so close to home. If matters regarding stars of zero magnitude are in this state, it cannot bode well for the presumed understanding of all the other stars of which our own Galaxy, at least, is composed.

Reference spectra like the Arcturus Atlas (1968) and the Procyon Atlas (1979) have been widely acclaimed and are still in demand, but remain the only ones of their kind for any star other than the Sun. Work of that size is a full-time effort, and both deserves and needs the support of the astronomical community which will benefit from it. In seeking to address this matter I would like to propose the establishment of a long-term international Center for Spectroscopic Studies, whose goal will be the generation, preparation and dissemination of very high-quality spectra of many of the naked-eye stars, together with relevant laboratory data.

Workshop on Databases for Galactic Structure
A. G. Davis Philip, Bernard Hauck and Arthur
R. Upgren, eds. p. 43 - 43 © 1993 L. Davis Press

R. Elizabeth Griffin
Intitute for Astronomy
Cambridge, U.K.

THE ASTROPHYSICS DATA SYSTEM: UNIFIED ACCESS TO DISTRIBUTED ASTRONOMICAL DATA AND SERVICES

Guenther Eichhorn

Smithsonian Astrophysical Observatory

ABSTRACT: The ADS is a client-server based data retrieval system. It provides unified access to astronomical and astrophysics data located at various data centers/nodes. The data are located at, and maintained by the nodes. This ensures that they remain where the expertise for these data is located. The users can select and retrieve subsets of these data in real time and transfer them to their local workstations for data analysis. This greatly facilitates multi-wavelength and interdisciplinary research. We currently provide access to 195 catalogs. The ADS also provides other services useful to astronomers. Among these are access to NED and SIMBAD and a server for searching through the NASA RECON abstract data base with about 125,000 astronomy and astrophysics abstracts.

1. PROJECT HISTORY

The Astrophysics Data System (ADS) is the result of the initiative of a group of astronomers started in 1988. They estimated the amount of data that NASA missions would produce for the coming decade and found that a new system would be necessary to handle this massive amount of data and make it accessible to the astronomical community (Fig. 1).

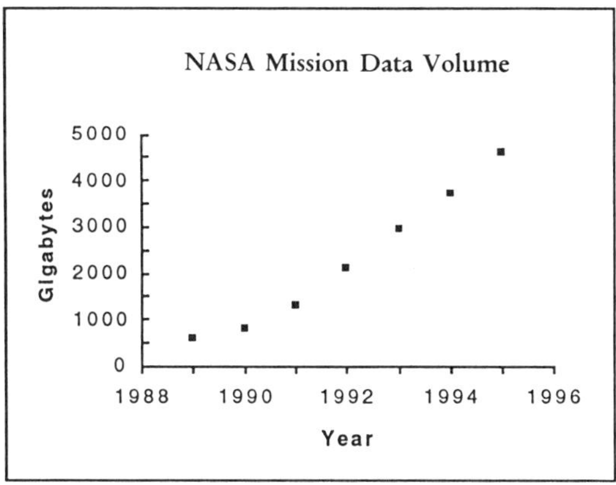

Fig. 1. Projected data volume collected by NASA space missions. This projection was done in 1988 and is by now known to be an underestimate.

Workshop on Databases for Galactic Structure
A. G. Davis Philip, Bernard Hauck and Arthur
R. Upgren, eds. p. 45 - 51 © 1993 L. Davis Press

Guenther Eichhorn
Center for Astrophysics
60 Garden Street
Cambridge, Massachusetts
02138

A workshop reported to NASA in 1988 about the conclusions they had reached and NASA approved a working group to define a data access system. The consensus in this group was that the amount and nature of the data required a distributed system. The data would be made available by the centers that collected them and accessed by the users through a distributed client/server system. This would make it possible to maintain the data and update them as necessary without having to re-distribute them after each update. Whenever users accessed these data, they would have the latest version available.

A prototype of a user interface and data access system was released in June 1990 and the initial operating system in June 1991. The user interface was a character based interface. Revision 2 of this interface was released in April 1992. It was soon realized that in order to make the system easy to use, a graphical user interface (GUI) had to be developed. The first version of the GUI was released in November 1992 and an update with added functionality in April 1993.

Currently ADS provides access to 195 catalogs and 125,000 astronomy and astrophysics abstracts. The next release, currently scheduled for October 1993, will provide access to several data archives (Einstein, IRAS, IUE, ROSAT, etc.) in addition to catalog access.

2. PROJECT PARTICIPANTS

The ADS depends strongly on the cooperation of many groups. There are two major parts of the ADS project: The development team and the data centers. Both of these are distributed throughout the US. The development team consists of:

Smithsonian Astrophysical Observatory, Harvard/Smithsonian CfA, Cambridge, MA
Infrared Processing and Analysis Center, Caltech/JPL, Pasadena, CA
Center for Astrophysics and Space Astronomy, University of Colorado, Boulder, CO
Ellery Systems, Inc.

The data centers are (in addition to the data centers at the development groups):

Department of Astronomy, Pennsylvania State University, University Park, PA
Space Telescope Science Institute, Baltimore, MD
National Space Science Data Center, Goddard Space Flight Center, Greenbelt, MD
International Ultraviolet Explorer Mission Center, GSFC, Greenbelt, MD
High Energy Astrophysics Science Archive Research Center, GSFC, Greenbelt, MD
GRO Science Support Center, Goddard Space Flight Center, Greenbelt, MD
Center for EUV Astrophysics, University of California, Berkeley, CA.

3. PROJECT SCOPE

Astronomical research relies more and more on the comparison of data collected with different instruments and their comparison with theoretical models. More and more complex questions require access to more data collected in more wavelength bands by more sophisticated instruments. Studies using these diverse wavelength bands will provide significant new science results if the data can be made accessible to a large user group. Most of the instruments collecting these data are generally not directly accessible to the astronomy researcher either because of oversubscription or because they are on space missions. With spacecraft such as HST, GRO, ROSAT, IUE, EUVE, COBE, etc. new areas of astronomy are open to the astronomical community. To be able to utilize this vast amount of multi-spectral data the researcher needs tools to conveniently access and analyze the archived data.

The ADS is such a tool. It is a distributed environment that provides the user access to a variety of astronomical data and services. At present the main emphasis is on making data collected by NASA space missions available to the astronomic community. However, this is presently being expanded to also include access to data from ground based observations.

The ADS provides access to a wide variety of data and services. It allows the users to retrieve selected data to their workstations and to do preliminary data manipulation and analysis. The user can specify sophisticated selection criteria to determine which data need to be retrieved for the research in progress to minimize the amount of storage required. This allows the user to retrieve data from many data sets concurrently without having to download all the complete data sets.

ADS is a tool to select and retrieve data. As such it provides utilities to visualize the data to make sure they are what is needed for the research and some basic data manipulation tools. However it is not a data analysis tool. For the in-depth data reduction other analysis systems like IRAF, AIPS, MIDAS should be used. The use of FITS I/O standards for the importing and exporting of data files makes it easy to transfer data between ADS and the data analysis tools.

4. SYSTEM ARCHITECTURE

The ADS is a distributed client/server system. The users are supplied by ADS with the client software that runs on their local workstations. This client software provides the graphical user interface and the networking software to access the servers at the data centers.

The KERBEROS authentication system is used to verify user passwords during login. This system provides the capability of setting up access control lists for services and secure authentication of users at the data centers through encrypted authentication distribution.

Data retrieval is done in several steps. First the user selects the data source and specifies the selection criteria. The client software then contacts a service called Locator. All services as they start up on the data center server register themselves with the Locator. When a client inquires about a specified service, the Locator returns the access information for that service. The client then contacts the service directly and establishes a connection. It then sends the request for data to the server and eventually receives the retrieved data.

The data returned are in table format for catalog queries and in FITS format for archive queries. ADS provides tools to manipulate and plot these tables and to display retrieved FITS images. Other data may be retrieved in different formats depending on the nature of the data. The ADS table manipulation tools include relational data base tools for working with multiple tables (selecting, merging, and correlating tables) and a table calculator for mathematical manipulations of tables.

Some services (e.g. plot or display services) are provided locally as well as remotely. Accessing a service locally is faster, but it requires more storage on the user site. We leave it up to the users whether they want to use services locally or remotely.

ADS is also available as a tool for manipulating local data sets that are not advertised to the network of ADS users. This may be required for recently acquired data sets that are still proprietary to the investigator that produced the data. This allows the use of the ADS data manipulation and display tools for the day-to-day activities in a research laboratory.

5. USING THE ADS

First a query to NED with the object name 'Coma Cluster' provides the coordinates of the Coma cluster. Then a query to the CfA Redshift Catalog is done to get a list of galaxies centered on the cluster within a 1° range in right ascension and declination with their classifications. The table editor is then used to create two tables, one for elliptical and one for spiral galaxies. Next, a request to the EINSTEIN archive provides an image of the Coma cluster in X-rays. This can be done by just specifying the name "Coma" as a search parameter. The image display tool then displays the X-ray image and plots the distribution of elliptical and spiral galaxies over it. It is clear from this image that the elliptical galaxies (circles) are concentrated in the center whereas the spiral galaxies (X's) are on the periphery of the cluster (Fig. 2). Finally a query to the abstract service with a natural language query asking for "Coma cluster, galaxy type classification and distribution" retrieves a list of abstracts sorted by their relevance to the scenario shown here (Fig. 3).

6. ADS STATUS

The ADS has currently about 1200 registered users. For the month of May, 300 distinct users logged onto the system a total of 4900 times. The number of queries for data in May was 5600, the total amount of data retrieved was on the order of 100 MBytes. About 18000 short abstract listings and 4000 full abstracts were retrieved in May. This usage is steadily increasing and is already surpassing some of the popular other services (e.g. SIMBAD).

ACKNOWLEDGEMENT

I wish to thank all the participants in the ADS for their support of the project.

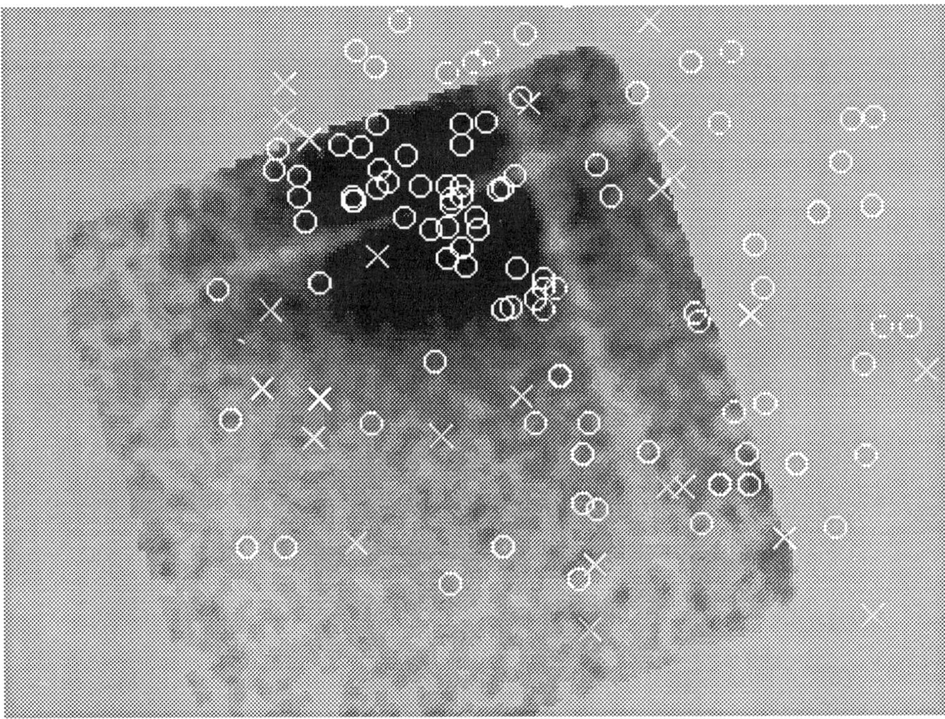

Fig. 2. Distribution of elliptical (circles) and spiral (crosses) galaxies overlayed over an x-ray image of the Coma Cluster of galaxies. The elliptical galaxies are clustered near the center of Coma whereas the spiral galaxies are distributed around the perimeter.

```
▣ Abstracts List Window
 File  Edit  View  Query

 # records returned:  [  10  ]              total # returned from query:  [ 54753 ]
```

Accno	Date	Score	Authors	Title
A90-41083	06/90	1	GAVAZZI, G.+GARILLI, B.+BOSELLI, A.	A CCD survey of galaxies in the Coma supercluster
A90-36765	00/88	1	FITCHETT, MICHAEL J.	Substructure in clusters of galaxies
A86-16509	11/85	1	ELLIS, R. S.+COUCH, W. J.+MACLAREN, I.+KOO, D. C.	Spectral energy distributions for galaxies high-redshift clusters. II - 0016 + 16 at 0.54
A77-28872	03/77	1	DAWE, J. A.+DICKENS, R. J.+PETERSON, B. A.	Redshifts of galaxies in the cluster HMS 1247-4102 in Centaurus
A88-47312	07/88	0.892	SLEZAK, E.+MARS, G.+BIJAOUI, A.+BALKOWSKI, C.+FONTANELLI, P.	Study of a field in the Coma supercluster - Automated galaxies count
A81-44806	07/81	0.892	SULLIVAN, W. T., III+BATES, B.+BOTHUN, G. D.+SCHOMMER, R. A.	Spiral galaxies in clusters. I - Neutral hydrogen observations in Abell 1367, Coma, and Zwicky 74-23 /1400.4 + 0949/
A81-21236	00/80	0.892	JONES, C.	The structure and evolution of X-ray clusters of galaxies
A87-38390	04/87	0.887	THOMPSON, LAIRD A.+VALDES,	High-resolution imaging from Mauna Kea -

```
 [2,2] /home/gei/tmp/ktIAAa01872 Row 11 of 26 Pos 21 of 191 T
 ◁
Select abstracts to retrieve:  Position cursor with left mouse button and Select with right mouse
```

Fig. 3. Results of an abstract query using the words "Coma Cluster, galaxy type classification and distribution".

DISCUSSION

PHILIP: Can the user assign the weights or is that a function only of the ADS?

EICHHORN: The user can assign weights for the calculation of combined scores in the abstract service as well as boolean combination methods. This sophisticated weighting scheme is unique to the ADS

RICHMOND: Could you describe the process a site must go through to become an ADS server?

EICHHORN: The degree of difficulty of bringing a node on line depends on the type of data and database used. To start a node with catalog access that uses one of the databases that ADS already supports (Ingres, Sybase, Allbase) takes only a day or two. For other databases a server program needs to be written. The complexity of this program depends on the complexity of the database query language, but in general this is an effort of a couple of weeks. For data archive access, the effort is higher. It depends on the structure of the archive and the extent of available documentation.

CASERTANO: The graphic interface looks very nice, easy to use, and powerful. Yet, as a potential user, I run into practical problems installing it. Wouldn't it be nice to have the option of a simple, text-based interface that would not require user installation?

EICHHORN: A character based interface would require almost as much installation since all the networking software needs to be installed. We would like to be able to support a character based interface, but funding restrictions prevent us from doing so at the present time.

GRIFFIN: In Europe, ESA and ASTRONET have put considerable resources into developing ESIS, which is currently at a stage of development that is fairly compatible with what you have been describing. Do you regard ESIS as a collaborator or a competitor?

EICHHORN: If ESIS would operate in the US it would be a competitor. As it is I would like to cooperate with ESIS and try to make our resources mutually available to users of both ADS and ESIS.

ROMAN: I got caught farther back. I do not have Windows. Hence, after I registered, I found I could not use the ADS.

EICHHORN: See answer to Casertano.

CASERTANO: Yes, I ran into Windows-related difficulties as well, with unresolvable or out-of-date widgets, if I recall correctly. If a simple, text-based interfere had been available, I would already have made significant use of the ADS system. As it is, I have put it off until such time when I can put together the disk space, my own time, and the system manager's assistance to install the system. I will also say that ADS people have been very quick and helpful in assisting me with my queries and difficulties, and that users will always ask for something more or - as is the case here - something less.

EICHHORN: The new version that was released on 1 April, 1993 has fixed most of the problems with x-widgets. If problems remain, please let us know.

GARCIA: How does ADS handle "version checking" between multiple servers with the same data set?

EICHHORN: This needs to be done by cooperation between the nodes that carry the identical data sets. In general ADS does not impose any restrictions on the data that a node supplies and does not check for validity or consistency. This is the responsibility of the nodes.

THE THIRD CATALOGUE OF NEARBY STARS: COMPLETENESS AND STELLAR KINEMATICS

Hartmut Jahreiss and Wilhelm Gliese[*]

Astronomisches Rechen-Institut Heidelberg

ABSTRACT: A direct investigation of space distribution and kinematics of low-mass stars can be carried out only in the immediate solar neighborhood. This fact emphasizes the importance of the Third Catalogue of Nearby Stars (CNS3) as a probe of the stellar content of our galaxy.

The CNS3 contains information on all known stars within 25 parsecs based on an extensive literature search during almost four decades. It profits now also, to a large extent, from recently published or updated compilations of astrometric, photometric and spectroscopic data. In addition to available basic properties - parallaxes, positions, proper motions, radial velocities, spectral types, photometry - further information on space velocities, duplicity, variability and other peculiarities is supplied as far as possible. During the course of the preparation advantages but also deficits of existing databases became obvious, and certain desiderata for an improvement will be brought up for discussion.

The completeness of the CNS3 with respect to various aspects and its consequences for the investigation of space distribution and kinematics will be discussed in more detail.

1. INTRODUCTION

In 1991 a preliminary version of the Third Catalogue of Nearby Stars was compiled to become available on the CD-Rom of the ADC (Gliese and Jahreiss, 1991). The first results on the content were published in Jahreiss and Gliese (1993) and Jahreiss (1993). To summarize: the preliminary CNS3 contains 3264 systems with altogether 3803 resolved components (not counted were the 195 spectroscopic binaries). The stellar number density drops from 0.11 stars per cubic parsec within five pc distance very steeply to 0.04 stars per cubic parsec for the distance interval from 20 pc to the catalogue limit of 25 pc. Yet, the decrease in stellar number density depends very strongly on the absolute magnitude of the stars in so far as the catalogue is complete for stars with absolute visual magnitudes brighter than about eight mag. In the present paper we want to illustrate some further properties in more detail.

2. COMPLETENESS OF DATA

Due to the observational history of the stars - 36 per cent were collected within the last decade - the coverage of the whole sample with respect to various astrometric, photometric or spectroscopic observations is rather inhomogeneous.

[*]Deceased, June 12, 1993

Workshop on Databases for Galactic Structure
A. G. Davis Philip, Bernard Hauck and Arthur
R. Upgren, eds. p. 53 - 63 © 1993 L. Davis Press

Hartmut Jahreiss
Astronomisches Rechen-Institut
Moenchhofstr 12 - 14
D-6900 Heidelberg 1, Germany

TABLE 1

Incompleteness in trigonometric parallaxes
and radial velocities

m	N(*)	no	%	no RV	%
6	550	31	6	0	0
7	238	47	20	36	15
8	292	76	26	61	21
9	262	48	18	44	17
10	340	67	20	99	29
11	353	134	38	217	61
12	361	178	49	278	77
13	381	244	64	323	85
14	258	166	64	224	87
15	130	52	40	121	93
16	65	22	34	59	91
17	19	6	32	15	79
18	15	0	0	10	67
all	3264	1071	33	1487	46

(*) only single stars or A-components

2.1 Parallaxes

The basic distance parameter is the trigonometric parallax. It was taken from a 1989 version - generously supplied to us by Prof. van Altena - of the forthcoming Yale Parallax Catalog (van Altena et al. 1993). In this version internal and external parallax errors were given, and we took the larger one as our parallax error. This is contrary to later versions of the YPC containing only remarks on the agreement of the measurements carried out at different observatories. Our different approach affects only the errors of a small percentage of the nearby stars. A discussion of the system of the trigonometric parallaxes in the YPC can be found in these proceedings (van Altena et al. 1993). Further information will be presented in Jahreiss (1993).

Trigonometric parallaxes were measured for 69 per cent of the CNS3 stars. Yet, for 28 per cent of these stars a "better" photometric parallax was chosen as resulting distance estimate. Altogether 50 per cent of the CNS3 parallaxes originate from photometric or spectroscopic measurements in combination with recently calibrated color-luminosity relations for dwarf stars (Jahreiss and Gliese, 1988).

The average parallax error is about 0.007 arcsec. But, still 19 per cent of the CNS3 stars have questionable parallaxes, in so far as trigonometric and photometric parallaxes are very discrepant.

Table 1 shows the number of all stars (column 2) with respect to visual magnitude (column 1). In column 3 the number of stars still missing trigonometric parallax measurements is given (the percentage is listed in column 4). It becomes evident that the amount of photometric parallaxes is especially high in the magnitude range from 11 to 15. Most of these objects entered the catalogue due to distance moduli from subsequent photometric measurements of high-proper motion stars.

2.2 Proper Motions

Almost all stars in CNS3 have measured proper motions. The main source (67 per cent) is the CDA (Jahreiss 1989) - a selection of the "best" positions and proper motions compiled from astrometric catalogues. The initial aim of the CDA was to supply the HIPPARCOS Input Catalogue stars with a required positional accuracy better than 1.5 arcsec for epoch 1991. Additional 25 per cent proper motions were taken from photographic surveys (with preference to Luyten's values). For the final version presumably 20 per cent of the stars will receive improved proper motions from PPM-south.

Table 2 presents a global comparison of proper motions determined by different methods. Columns 2 to 4 give $\Delta\mu_\alpha$, $\Delta\mu_\delta$, and $\Delta\mu$ in arcsec/yr revealing possible systematic differences; whereas the corresponding absolute differences - showing the scatter of the catalogues - can be found in columns five to seven. Table 2 lists the median values instead of means to suppress the influence of possible outliers. At first, recently determined proper motions using Astrographic Catalogue positions and modern meridian observations (Jahreiss et al. 1992) were compared with Luyten's photographically measured NLTT values. Secondly, relative proper motions from the Van Vleck trigonometric parallax observations (VVO stars) were compared with new-AC and PPM proper motions.

The average error of the new-AC proper motions amounts to about 0."004/yr. This leads to a scatter of the NLTT values in the order of 0."015/yr in accordance with Luyten's error estimate of 0."014/yr for his machine measured proper motions. No systematic differences with respect to the FK4 system are significant. On the other hand, the scatter of the VVO proper motions is smaller, but, a significant systematic deviation is present.

Yet, we have to concede that this result applies only to stars brighter than the AC limit (m_V = 12 mag), and the NLTT data may still contain some SAO values. In the most recent USNO parallax list Harrington et al. (1993) find no systematic differences between their relative proper motions and NLTT, and their sample contains mostly much fainter objects. These findings justifies the use of NLTT proper motions unless the more accurate proper motions from parallax series were reduced to absolute.

2.3 Radial Velocities

During recent years several extensive lists of photoelectrically determined radial velocities were published by various authors measured with different instruments. In the absence of a uniform system - the discussion about a correct zero point especially for late-type K and M

TABLE 2

Comparison of various proper motion sources

	$\Delta\mu\alpha$	$\Delta\mu\delta$	$\Delta\mu$	$\mid\Delta\mu\alpha\mid$	$\mid\Delta\mu\delta\mid$	$\mid\Delta\mu\mid$
976 NLTT-stars: new-AC (French zone) versus NLTT						
	+0.001	-0.003	-0.001	0.015	0.016	0.017
57 VVO-stars : new-AC versus VVO						
	+0.003	+0.011	-0.013	0.009	0.012	0.014
140 VVO-stars : PPM versus VVO						
	+0.006	+0.010	-0.009	0.009	0.012	0.012

stars is going on - the available measurements for such stars were reduced to the system of Marcy and Benitz (1989). For almost 1000 CNS3 stars photoelectric radial velocities were available. Typical errors of these measurements should be in the order of 1 km/sec, usually for those stars existing spectroscopic values having much larger errors were not taken into account.

According to Table 1 (columns 5 and 6) the lack of radial velocities becomes very dominant for stars fainter than m = 10. This certainly affects the study of the kinematics for the very low-mass stars. A KS-test shows that missing radial velocities (about 15 per cent) for the nearby K- and M-dwarfs found by Vyssotsky and collaborators in a spectroscopic survey (m_V = 7 - 11 mag) should produce no kinematical bias.

Within a few years the situation may improve considerably. Then for example the CORAVEL measurements for K and M dwarfs will become available and the ESO Key Program on radial velocities for Hipparcos stars will be finished (Mayor et al. 1992). The portion of nearby stars without radial velocities should then become smaller than 15 per cent down to m_V = 12 mag.

3. SPATIAL DISTRIBUTION

The stellar number density is rather constant within the catalogue limit of 25 pc for the absolutely brighter stars. Yet, two remarkable features show up (see Fig. 1) in plotting the normalized cumulative stellar number density of the stars brighter than $M_V = 8$ mag versus distance.

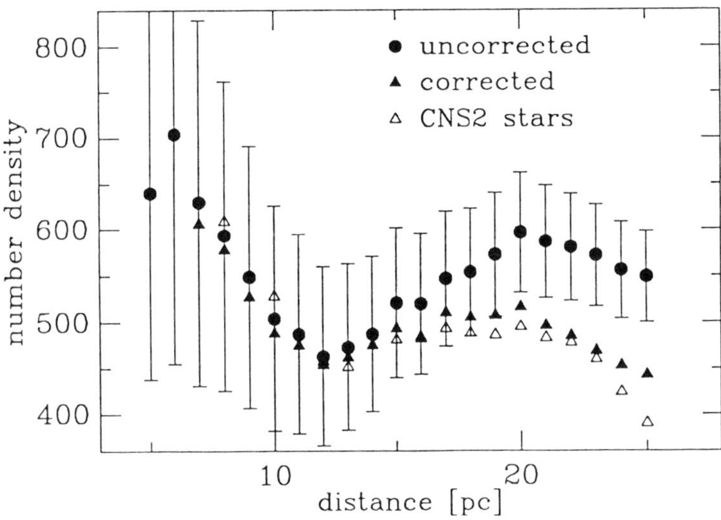

Fig. 1. Cumulative number density of CNS3 stars with $M_V < 8$ mag normalized to a volume within 20 pc. Filled circles: original data with Poisson errors. Filled triangles: after correction. Open triangles: CNS2 stars for comparison.

There is a distinct maximum at 20 pc - the former catalogue limit. Since we are dealing with a volume-limited sample more stars with positive parallax errors enter the catalogue, and the fact that the parallax errors of the brighter stars are quite large strengthen this effect. To correct for it, poor trigonometric parallaxes of non-dwarf stars were replaced by photometric ones with the result that distances became larger and the maximum at 20 pc for the number density smaller (Fig. 1 filled triangles). Yet, the maximum is still present, and it cannot be explained by the way the data were binned. So, it must be assumed that there is very probably a selection effect, which should be investigated in more detail before the final version of the CNS3 will be finished. For comparison, the CNS2 stars (open triangles) were plotted, indicating that already in 1969 the desired completeness limit of 22 pc could be achieved.

A second feature becomes conspicuous in Fig. 1. There seems to be a density dip at a distance of about 12 pc. Of course, the number of stars involved is quite small, and various other approaches couldn't make it more conclusive. So, we simulated uniform spatial distributions of a corresponding number of stars within 25 pc. Fig. 2 shows one example of such a distribution. It also reveals a large scatter. Taking the mean of 1000 runs (filled circles) produces the expected value for almost the complete distance range. Whereas the median values (open circles) indicate that for smaller distances (r < 13 pc) there occur more deviations from the expected values to smaller densities than the other way.

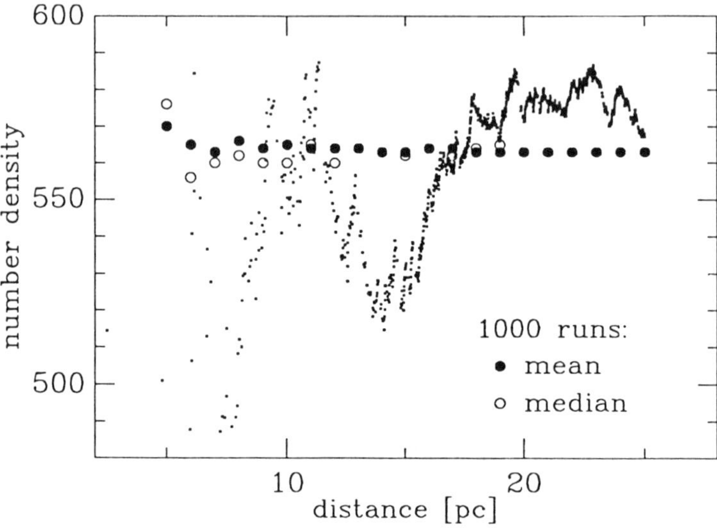

Fig 2 Cumulative number density of 1110 simulated stars (dots) uniformly distributed within 25 pc. Filled circles: mean of 1000 runs. Open circles: median of 1000 runs.

For the absolutely fainter stars ($M_v > 11$ mag) the luminosity function as determined by Wielen et al. (1983) depended mostly on stars within 5 pc. The number of known objects within this limit increased from 52 stars in 1957 (CNS1) to 54 stars in 1969 (CNS2) and to 66 stars in 1991 (CNS3). 70 stellar objects within 5 pc of the Sun are presently known if we count also reliable astrometric or spectroscopic binaries, and it is very probable that additional ones will be detected in the near future. Take, for example, LHS 2924 - the CNS3 star with faintest absolute visual magnitude ($M_v = 19.63$), at a distance of 10.5 pc it is far beyond this limit.

Fig. 3 illustrates very nicely the rapid change in distance from the Sun with time due to the movement of the stars. For all CNS3-stars with known space velocity components the minimal distances from the Sun were calculated, and those stars were plotted in Fig. 3 with encounters smaller than two pc. It can be seen that within 40,000 years from now there is a clustering of several stars within one and two pc.

4. KINEMATICS

Additional information is necessary to carry out kinematical investigations (Jahreiss and Wielen 1983). For example individual ages of the stars are required or at least mean ages of homogeneous groups of stars to study the age dependence of the velocity distribution.

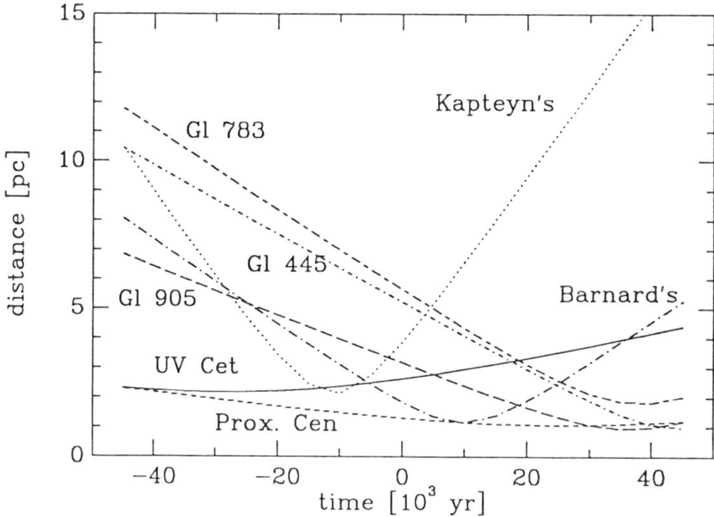

Fig. 3. The distance between the Sun and some nearby stars as a function of time from now.

Information on ages of stars more massive than the Sun can be determined from their position in the HR diagram. A separation of the stars on or near the main sequence in different (B-V) ranges for (B-V) < 0.65 produces different age groups. Table 3 presents such groups with the number of stars in column 2. The velocity dispersions σ_U, σ_V, σ_W and σ_{tot} in km/sec, and the axial ratio σ_V/σ_U are given in columns 3 to 7, and the mean age of each group (in 10^3 yrs) is listed in column 8. The first part repeats the results for the CNS2-stars within 20 pc. Since an extension from 20 pc to 25 pc yields twice the volume, the number of stars should be doubled if the catalogue is complete, which is the case for this region of the HR diagram. The second part of Table 3 (containing the original CNS3 data) shows indeed many more stars, especially in the very early groups. But, after correction of the CNS3 distances, as described in chapter three, the size of the groups is again reduced (see the third part), and reflects the "correct" 1:1 portion of stars within 20 pc or beyond. In other words, the distances of some early type stars were underestimated in CNS2. In view of this fact, it is quite astonishing, that there are no really significant differences in the resulting velocity dispersions for the different samples. This, presumably, is due to the small proper motions of the stars in question. Therefore, a change in distance doesn't affect the value of the space velocity components very much.

The last two parts of Table 3 compare the old (CNS2) and new (CNS3) values of the velocity dispersion for the 317 Vyssotsky stars and their subsamples selected according to O. C. Wilson's estimates of the CA II H+K emission line intensity. Since identical samples are compared the expected changes should be merely due to improved distances, proper motions and radial velocities. Again, no really significant differences can be seen, considering an error

TABLE 3

Velocity dispersion for different age groups
$|W|$-weighted

group	n	σ_U	σ_V	σ_W	σ_{tot}	σ_V/σ_U	$<\tau>$
CNS2 (r 20 pc)							
HR 6d	6	14	8	3	16	0.53	0.2
HR 6c	14	20	7	4	22	0.34	0.5
HR 6b	16	15	12	8	21	0.80	1.0
HR 6a	47	31	20	11	39	0.64	2.3
HR 5	83	42	26	25	56	0.62	5.0
CNS3 (r 25 pc) uncorrected							
HR 6d	16	13	7	6	16	0.59	0.2
HR 6c	29	21	7	7	24	0.34	0.5
HR 6b	35	19	14	10	26	0.73	1.0
HR 6a	94	24	15	13	31	0.62	2.3
HR 5	173	39	27	23	52	0.69	5.0
CNS3 (r 25 pc) corrected							
HR 6d	8	14	8	2	16	0.53	0.2
HR 6c	15	21	7	8	24	0.35	0.5
HR 6b	21	14	11	8	19	0.80	1.0
HR 6a	79	24	16	14	32	0.63	2.3
HR 5	156	40	27	23	53	0.69	5.0
Vyssotsky stars CNS2-sample, old values							
all	317	48	29	25	62	0.60	5.0
HK +8,+3	23	20	10	6	23	0.50	0.3
HK +2	40	22	17	13	31	0.75	1.4
HK +1	36	30	16	15	37	0.55	3.0
HK 0	41	40	21	20	50	0.52	5.2
HK -1	24	40	34	34	63	0.86	7.2
HK -2,-5	31	67	29	25	77	0.44	9.0
Vyssotsky stars CNS2-sample, new values							
all	317	49	33	25	64	0.67	5.0
HK +8,+3	23	21	10	7	24	0.47	0.3
HK +2	40	24	17	14	33	0.71	1.4
HK +1	36	31	17	16	38	0.56	3.2
HK 0	41	41	23	18	50	0.57	0.3
HK -1	24	47	37	35	69	0.79	7.2
HK -2,-5	31	68	28	25	78	0.42	9.0

can be seen, considering an error of three km/sec for the resulting velocity dispersions. The same is true if the velocity dispersion of the 353 Vyssotsky stars within 25 pc are analyzed.

The comparisons in Table 3 demonstrate that already now quite reliable results can be obtained. Nevertheless, it is highly desirable to get further progress especially for the low-mass stars. At present, O. C. Wilson's HK estimates from 1970 are still the most comprehensive parameters allowing at least a relative age separation for low-mass stars. In the meantime there exist devices to measure the HK flux relative to the nearby continuum, and extensive observing runs were already carried out. But, to get the true chromospheric emission it is necessary to correct for the variation of the photospheric flux with spectral type. This is, presently, not feasible for stars later than K5, i.e. (B-V) > 1.00 (see for example Soderblom 1985). So, additional observations and investigations are required to get more reliable age determinations for the low-mass stars.

5. CONCLUSIONS

Without any doubt, the CNS3 is a very useful tool to investigate space distribution and kinematics in the immediate solar neighborhood. Yet, it is no perfect tool. In applying it, one has to keep in mind that the CNS3 is incomplete in several aspects, and one has to use carefully selected subsamples to avoid selection effects that may falsify the desired results.

A considerable improvement of this tool will be achieved by the Hipparcos satellite and related observing programs. Within several years not only much better astrometric parameters will become available for most nearby stars brighter than m_V = 12 mag, but also very useful information on duplicity and variability as well as further spectroscopic and photometric data.

Among the absolutely fainter stars those found in high-proper motion surveys are presently over-represented. To get more information on the low-velocity tail, projects as for example DENIS (Epchtein, 1993) or the identification of nearby X-ray sources detected by ROSAT in other wavelength look very promising.

REFERENCES

Epchtein, N. 1993 in Workshop on Database for Galactic Structure, A. G. D. Philip, B. Hauck and A. R. Upgren, eds, L. Davis Press, p. 97
Harrington, R. S., Dahn, C. C., Kallarakal, V. V. et al. 1993 AJ 105, 1571
Gliese, W. and Jahreiss, H. 1991 on The Astronomical Data Center CD-ROM: Selected Astronomical Catalogs, Vol. 1, L. E. Brotzmann and S. E. Gesser, eds., NASA/ADC, Greenbelt
Jahreiss, H. 1989 in The HIPPARCOS Mission, Pre-Launch Status, Vol II: The Input Catalogue. M. A. C. Perryman (Scientific Coordinator), European Space Agency, Paris, ESA SP-1111, p. 115
Jahreiss, H. 1992 in HIPPARCOS, Comptes-Rendus de l'Ecole de Goutelas 1992 D. Benest and C. Froeschle, eds., p. 593
Jahreiss, H., Crifo, F. and Requieme, Y. 1992 in Highlights in Astronomy, Vol. 9, J. Bergeron, ed., Kluwer Academic Press, Dordrecht, p. 399
Jahreiss, H. and Gliese, W. 1993 in IAU Symposium No. 156, Developments in Astrometry

and Their Impact on Astrophysics and Geodynamics, I. I. Mueller and B. Kolaczek, eds., Kluwer Academic Press, p. 107

Jahreiss, H. and Wielen, R. 1983 in IAU Colloquium No. 76, The Nearby Stars and the Stellar Luminosity Function, A. G. D. Philip and A. R. Upgren, eds., L. Davis Press, Schenectady, p. 277

Marcy G. W. and Benitz K. J. 1989 ApJ 344, 441

Mayor, M., Gerbaldi, M., Grenier, S. and Levato, H. 1992 in Highlights of Astronomy, Vol. 9, J. Bergeron, ed., Kluwer Acadademic Press, p. 433

Soderblom, D. R. 1985 AJ 90, 2103

van Altena, W. F., Lee, J. T. and Hoffleit, E. D. 1993 in Workshop for Databases in Galactic Structure, A. G. D. Philip, B. Hauck and A. R. Upgren, eds., L. Davis Press, Schenectady, p. 65

Wielen, R., Jahreiss, H. and Krüger, R. 1983 in IAU Colloquium No. 76, The Nearby Stars and the Stellar Luminosity Function, A. G. D. Philip and A. R. Upgren, eds., L. Davis Press, Schenectady, p. 163

DISCUSSION

CASERTANO: Magnitude systems based on low-accuracy trigonometric parallaxes (error 10 - 15%) are notoriously biased and ill-defined. Have you applied some kind of correction to take such errors into account? As an alternative, better-behaved test of uniformity of the sample, have you considered studying the distribution of the trigonometric parallaxes themselves?

JAHREISS: At the moment my corrections depend merely on replacing poor trigonometric parallaxes by better photometric ones. The distribution of parallaxes was studied only for fainter stars.

UPGREN: A few years ago I made an investigation of the distribution of parallaxes with distance. For CNS2 and the 1979 supplement, it fit a power law of π^{-4} (which holds for a uniform distribution in transparent space). For stars with $M_v < +9$, the -4 law held to the distance limit of 22 pc, but for stars with $M_v > +9$ (spectral type about M0 V), this distance limit quickly "moved in" to about 13 pc. This indicated completeness to ~ 13 pc but great incompleteness for dM stars beyond that distance.

CREZE: What are the main sources for the detection of new nearby stars between the two versions of the catalogue?

JAHREISS: Subsequent photometric and trigonometric observations of proper-motion survey stars.

HUMPHREYS: I was surprised that you gave a survey limit of $m_v \sim 21$ mag from Luyten's proper motion. Based on re-scanning some of his plates, we find his data on motions become incomplete beginning at ~ 17.5 - 18 mag on the E (red) plate.

JAHREISS: I took it from Luyten, but I didn't regard it as a completeness limit. In any case we know that the distribution in apparent magnitude declines well before the limit at about $m_{pg} = 17$.

CREZE: You mentioned 50 stars without proper motions. Where do they come from?

JAHREISS: They came from spectroscopic surveys, as for example Stephano's.

UPGREN: How much is the distribution of percentage of stars with radial velocities as a function of magnitude (especially the faintest stars) due to faint companions of brighter stars?

JAHREISS: In the statistics I presented only single stars and A-components were taken into account.

THE NEW EDITION OF THE YALE PARALLAX CATALOGUE AND ITS WEIGHTING SYSTEM

W. F. van Altena, J. T. Lee and E. D. Hoffleit

Yale University Observatory

ABSTRACT: The new edition of the General Catalogue of Trigonometric Parallaxes contains 15,430 determinations of parallaxes for 7,888 stars. Auxiliary data including cross-identifications, positions, photometry, spectral types and luminosity classes, and data on the duplicity and variability of the stars are given for each star. In addition, a reference to the publication for each parallax is given so that users can examine the original paper to extract further information. In this paper we describe the compilation of the data and the comparison of parallaxes determined at different observatories that resulted in the adopted system of parallaxes and their errors.

1. INTRODUCTION

In earlier publications (van Altena 1986, van Altena and Lee 1988, van Altena, Lee and Hoffleit 1991a) we described our work leading up to the publication of the CD-ROM "The General Catalogue of Trigonometric Parallaxes, Preliminary Version, 1991" (van Altena, Lee and Hoffleit 1991b). In this paper we will briefly review that work and define the weighting system adopted for the General Catalogue of Trigonometric Parallaxes, Fourth Edition 1993. The issues and data relating to systematic differences between parallaxes determined at various observatories are discussed in van Altena, Lee and Hoffleit (1993a).

2. DATA BANK

The data bank for the new edition of the YPC was generated by three independent literature searches for publications listing trigonometric parallaxes which resulted in a bibliography of approximately 500 citations. From these publications we extracted the relative parallax and proper motion, right ascension and declination, mean magnitude of the reference stars, reduced magnitude of the parallax star, cross-identifications, data on the star about its duplicity, variability and other interesting characteristics. Using the positions and cross-identifications, we then searched about 50 catalogues for improved positions and proper motions, UBV photoelectric photometry and MK Spectral Types and Luminosity Classes. Finally, additional catalogues were searched to improve the completeness of our cross-identifications. Quality assurance has been enforced by proofreading, computer searches for obvious blunders and making prepublication versions of the YPC available to the astronomical community for its use, and help in detecting errors.

Workshop on Databases for Galactic Structure
A. G. Davis Philip, Bernard Hauck and Arthur
R. Upgren, eds. p. 65 - 70 © 1993 L. Davis Press

William F. van Altena
Yale University Observatory
New Haven, Connecticut
06511

3. WEIGHTING SYSTEM OF THE YPC

Some of the stars in the YPC have more than one measurement; it was therefore necessary to derive a weighting system for the combination of the data into a most probable value for the parallax. However, since the mean magnitude of the reference stars varies between observatories and also evolves with time at each observatory as the individual program goals change, it was first necessary to correct each relative parallax to absolute. For that purpose, advantage was taken of the renewed interest in galactic modeling and new corrections to absolute calculated for each star using a three component model consisting of a disk, thick disk and a spheroid that represented the best estimates at that time for the distribution of the constituents of our galaxy. The corrections to absolute parallax computed with that model are discussed by van Altena, Lee, Hanson and Lutz (1988) and will be published in the near future.

The errors of the parallaxes published with each determination represent the internal scatter of the observations but are generally thought not to represent the true error of the parallax determination. We designate those errors as the internal and external errors, respectively. Since there are no absolute standards with which to compare the parallaxes, we must examine different methods of evaluating the external errors and look for consensus among the methods. For this new edition, we have used three independent methods to evaluate the external errors: the Hertzsprung (1952) method; the Observatory Pairs method (Vasilevskis 1966); and the Spectrophotometric parallax method. We have already discussed the first two methods in previous publications (see for example: van Altena 1986) and will therefore discuss only the last one here.

For the method of Spectrophotometric parallaxes, we rely on the fact that for distant stars the value of the derived parallax is much more precisely known than the trigonometric parallax for the same star. The difference between the two parallaxes for the same star is dominated by the error in the trigonometric parallax and it is then possible to compare the derived external error with the published internal error. Since the Spectrophotometric parallaxes were calibrated using trigonometric parallaxes it is necessary to consider the possibility of circularity in our comparisons. However, since the Spectrophotometric parallaxes were calibrated using only large parallaxes and we are restricting ourselves to small parallaxes (see the next paragraph) there is little danger of circularity in our analyses. Using the YPC data bank, with the assistance of G. K. Torrington, we derived external error estimates for all observatories with sufficient numbers of stars with UBV photometry and MK Spectral Types and Luminosity Classes.

Each star was first corrected for interstellar reddening using the intrinsic color and luminosity calibrations (UBV: Schmidt-Kaler 1982; VRI: Weis 1986, 1987, 1988, and Upgren (1993) for the MK Spectral Type and Luminosity Class and the UBVRI photometry. Differences between the trigonometric and photometric parallax were computed and Probability Plots used to estimate the dispersion of the differences for each observatory or observatory subset. By restricting the photometric parallaxes to some maximum value we are able to limit the contribution of the error in the photometric parallax to a negligible amount consistent with maintaining a reasonable number of stars in the sample. For most cases that limit was set at $0\overset{''}{.}015$, which for a cosmic dispersion of 0.6 mag. in the luminosities at a given spectral type results in an error in the derived photometric parallax of $0\overset{''}{.}004$ which is small in the context of the average trigonometric parallax error of $0\overset{''}{.}015$. The derived dispersion of the differences was then corrected for the inferred error of the photometric parallaxes to yield an estimate for the external error of the parallax of the sample. (For the cases of the newer parallaxes with internal errors comparable to 0".004, there are generally too few MK Spectral Types and Luminosity Classes, the existing calibrations are too poorly defined, or the calibra-

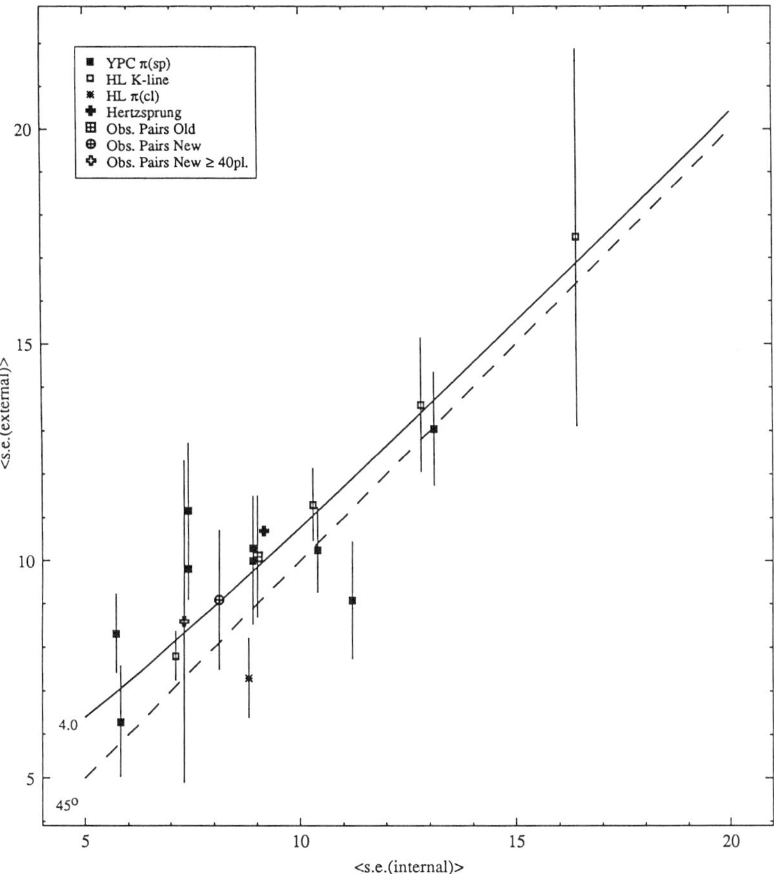

Fig. 1: Estimates for the external standard error of an Allegheny Observatory parallax versus the published internal standard error. Included are values derived at Yale from the Observatory Pairs analyses of both the old and new Allegheny parallaxes, all Allegheny parallaxes with more than 40 plates, values from the Spectrophotometric parallax analyses and from the Hertzsprung method. In addition, values are included from the Hanson and Lutz (1983) analyses of K-line spectroscopic parallaxes and from cluster parallaxes. The solid line represents the relation adopted for Allegheny Observatory parallax errors using Eqn. 1 with the scaling factor a = 1.0 and the constant noise b = 4.0 mas, while the dashed line represents equality between the external and internal errors.

tion stars dominate the sample and hence it is not possible to derive reliable external errors by this approach.)

The goal of this analysis is to derive the external error of the trigonometric parallaxes, since there is evidence that the published internal errors are only partially representative of the parallax errors. For the 1991 CD-ROM Preliminary Version of the YPC we based our analyses on the Hertzsprung and Observatory Pairs methods. From the data available at that time, we adopted an external error 15% larger than the internal error, except for parallaxes based on more than 40 plates, where the data seemed to indicate that the external and internal errors agreed. Now, with the addition of the estimates based on the spectrophotometric parallaxes, we have decided that a more appropriate relation between the external and internal errors is the quadrature sum of the scaled, a, internal error and some irreducible observatory error, b,

$$\sigma_{ext}^2 = \sigma_{int}^2 + b^2. \tag{1}$$

The validity of this model is of course open to question, but we have adopted it since it seems to represent the data for cases where an adequate number of error estimates is available.

In Fig. 1, we plot the various estimates for the external and internal errors derived for trigonometric parallaxes determined at the internal errors derived for trigonometric parallaxes determined at the Allegheny Observatory. In addition to our estimates of the external errors, we have also included the external error estimates by Hanson and Lutz (1983) derived from the K-line spectroscopic parallaxes and cluster parallaxes. The data are seen to be reasonably fit by Eqn. 1 with the constants a = 1.0 and b = 4 mas (1 mas = 0.001"). For each observatory with sufficient data the constants a and b were determined and they are listed in the Introduction to the YPC 1993. As a general rule, after looking at the comparisons, we have adopted a scaling factor of unity unless there is ample evidence to indicate otherwise. For only the Cape and McCormick observatories is there compelling evidence that the scaling factor is not unity and we have adopted zero for those cases, i.e. their parallax errors are best represented by a constant.

The issue of zero-point differences between the parallaxes determined at different observatories is discussed in van Altena, Lee and Hoffleit (1993a); they recommend that none be applied, except for the Allegheny and Dearborn magnitude-dependent differences.

ACKNOWLEDGMENTS

We would like to thank Mr. G. K. Torrington for his dedicated assistance in carrying out most of the calculations related to the determination of the errors derived from the Spectrophotometric parallaxes. This research has been supported in part by grants from the National Science Foundation.

REFERENCES

Hanson, R. B., and Lutz, T. E. 1983 MNRAS 202, 201
Hertzsprung, E. 1952 Observatory 72, 242
Schmidt-Kaler, Th. 1982 Landolt-Bornstein New Series Group 6, Vol. 2b Astronomy and

Astrophysics, Stars and Star Clusters, K. Schaifers and H. H. Voigt, eds., p. 14

Upgren, A. R. 1993, private communication

van Altena, W. F. 1986 in Astrometric Techniques, H. K. Eichhorn and R. J. Leacock, eds., Reidel, Dordrecht, p. 183

van Altena, W. F., and Lee, J. T. 1988 in IAU Symposium 133, Mapping the Sky, S. Debarbet, J. A. Eddy, H. K. Eichhorn and A. R. Upgren, eds., Reidel, Dordrecht, p. 269

van Altena, W. F., Lee, J. T., Hanson, R. B. and Lutz, T. E. 1988 in Calibration of Stellar Ages, edited by A. G. D. Philip, L. Davis Press, Schenectady, p. 175

van Altena, W. F., Lee, J. T. and Hoffleit, E. D. 1991a abstract of a talk given at the XXIst General Assembly of the IAU, in Transactions of the IAU XXIB, Buenos Aires 1991, J. Bergeron, ed., Kluwer, Dordrecht, p. 235

van Altena, W. F., Lee, J. T. and Hoffleit, E. D. 1991b The General Catalogue of Trigonometric Parallaxes, Preliminary Version, 1991b in Astronomical Data Center CD-ROM Selected Astronomical Catalogs, Volume I, L. E. Brotzman, S. E. Gessner, J. M. Mead and M. E. Van Steenberg, eds., Goddard Space Flight Center, Greenbelt van Altena,

van Altena, W. F., Lee, J. T. and Hoffleit, E. D. 1993a in Galactic and Solar System Optical Astrometry: Observation and Application, L. V. Morrison, ed., (in press)

van Altena, W. F., Lee, J. T., and Hoffleit, E. D. 1993b The General Catalogue of Trigonometric Parallaxes, Fourth Edition, Yale University Observatory, New Haven.

Vasilevskis, S. 1966 Ann. Rev. A&A, 4, 57

Weis, E. W. 1986 AJ 91, 626

Weis, E. W. 1987 AJ 93, 451

Weis, E. W. 1988 AJ 96, 1710

DISCUSSION

CASERTANO: Two comments, one pessimistic and one optimistic. The pessimistic one: even with Hipparcos, about 40% of the stars will fail the Lutz-Kelker test ($\sigma > 0.175\ \pi$). The optimistic one: there is life after Lutz-Kelker. If the question is asked correctly, especially for calibration purposes, parallaxes even less accurate than the L-K cutoff can be used to good effect.

CREZE: Hanson's work and Lutz and Kelker's corrections are not the last word about statistical use of parallaxes. There are methods to use low-accuracy parallaxes and some have been used successfully, although systematic errors may create difficulties.

UPGREN: The unfortunate situation that only about 1500 stars or about 20% have $\sigma\pi/\pi <$ 0.175, the Lutz-Kelker limit is at least a great improvement over the 3rd edition of the YPC and its 1963 Supplement where only about 4% were of this quality. The ground-based parallax programs have, in recent years, been observing the stars that they should be observing.

VAN ALTENA: I still urge that great care be taken when analyzing data that are so close to their errors. Systematic errors can dominate the parallaxes and the interpretation of those small parallaxes is perilous.

DISCUSSION I. (MODERATOR: LASKER) How Can We Improve Access to Data and Data Centers?

MODERATOR: The goal for this session is to consider how we can improve access to the data? We begin by noting that the data are found in several kinds of places, (a) in the (often relatively personal) archives of individual investigators, (b) in published databases, (c) in network-accessed and distributed databases, and (d) in the Data Centers. This of course leads us to still another question, i.e., before we can access data, we need to be able to determine where it is.

The discussion was assembled from hand-written comment sheets, then edited by the moderator, who takes full responsibility for any misinterpretations that may have been introduced in this process. A first discussion topic addresses the identity of the data, its descriptors, and its processing.

GRIFFIN: We should first define the term *data* carefully. The catalogues of derived quantities I call *information*, whereas the actual observations I will refer to as *raw* data. The term *catalogue*, is sometimes used to refer to indices of raw data. The distinctions are important because the treatment of and access to the relevant archives can be different.

ROESER: What kind of data should be stored at which data centers? Someone must address the matters of which data are *original* observational data, as opposed to *compiled* data, and what are the precise definitions of each.

ROMAN: If we are going to archive raw data, we must know what we mean by raw data. At a minimum, we must archive all of the instrumental characteristics. However, some reduction may be desirable. For example, for classification spectra, do we want to archive the original spectra? These might be useful for other purposes.

PHILIP: Probably we need two levels of data in databases. For example, in photometry, there are a lot of data published in which a set of photometric measures are published with no other details; one takes these data as is. However, in the case of CCD photometry, a deeper understanding requires that we also archive the data frames, bias, flats, dark, as well as the reduction algorithms. The important thing is to know, for any database, precisely in what state the data exist.

EICHHORN: I suggest that, together with data, we also archive the reduction algorithms and constants used in the data reduction. This will allow reprocessing in case a change to an algorithm or a constant becomes necessary.

URBAN: It can be very dangerous to present only raw data and algorithms, i.e., without corresponding catalogs. An example is the Astrographic Catalogue, whose data is "raw" (x-y coordinates and plate constants). This has hampered people using it, since few stars have spherical coordinates. When users want star positions, they want spherical coordinates. They do not want to play with algorithms that, eventually, if applied correctly, will yield spherical coordinates.

Workshop on Databases for Galactic Structure
A. G. Davis Philip, Bernard Hauck and Arthur
R. Upgren, eds. p. 71 - 75 © 1993 L. Davis Press

MODERATOR: The next discussion segment elaborates on some more subtle points associated with maximizing scientific understanding of the data.

CREZE: One of the most difficult aspects concerning the access to data is not related to database or network systems but rather to scientific specialization: the knowledge required to properly use observations from a space mission or from a modern ground-based system is getting so complex that non-specialists are unlikely to do it properly. There should be calibrated summaries prepared by specialists as part of their own investigation and then made available widely through data centers. This is more or less taken care of in the case of large surveys and for most space missions. But how about other investigations?

LASKER: These kinds of supporting materials are also increasingly available from the larger ground-based observatories.

RATNATUNGA: As a catalog end-user, what has taken me the most time is not obtaining the catalog but understanding its contents and quality. An end-user will not know the limitations, sampling bias, etc., for a catalog. I think the data centers should play a leadership role in making sure that the catalog and the full documentation and some basic statistical analysis can be made available to the general user. For example, many astronomical catalogs are published wherein some quantities have neither error estimates nor quality discussions; this is particularly troublesome many years after the data are taken.

EICHHORN: I feel it is important that the data producer provide good documentation with the data. The data center usually does not have the necessary expertise to do that with accuracy.

WARREN: We must face the fact that many researchers who compile data for a specific project will not be willing to take time to produce supporting documentation. There should be personnel at the data centers who have technical expertise and who are willing to interact with data suppliers to produce supporting information and to ensure that the data are properly archived.

CREZE: The only way around the problem of the quality control of catalog data is the refereeing. We at CDS are trying to get catalogues or data tables refereed as part of the papers that describe them. Data centers alone cannot perform scientific control.

RICHMOND: How can we allow users to create their *own* catalogs of information, based on their own software searches of "new" archived images and/or spectra?

LASKER: It seems that this capability is beginning to be provided by the several reduction/analysis packages that are available in the community. IRAF/SDAS is the one I know best, but there are others too.

CASERTANO: I would like to point out that, for the reduction of radio data, AIPS provides the perfect mechanism to keep track of the manipulation the data have gone through. All tasks *automatically* leave a record of their operation in the FITS header. This does not depend on the user's ability to keep things organized. Do other data analysis systems provide comparable functionality? If not, why not?

LASKER: Again, I note that at least some of the optical packages in common use do provide similar capabilities.

EICHHORN: Some data analysis and archiving systems discard the comments in FITS headers; this practice should be changed. Every system that handles FITS headers should preserve the complete reduction history in the FITS header; discarding FITS comments should be made a crime!

MODERATOR: The next question returns us to one of the major topics, how do we determine what data exist and where they are?

HAUCK: Is a directory of databases necessary or is the ADS the good answer to this question?

ROMAN: At present, the NSSDC Master Directory (MD) is a directory of catalogs which may or may not be in the NSSDC. The MD points to where these data are located. At present, the MD is not very well populated in astronomy, other than planetary. We will welcome information about data sets which we should catalog in the Master Directory. We also welcome catalogs for archiving but some data sets may be too large or otherwise unsuitable for archiving.

EICHHORN: ADS invites the astronomy community to provide catalogs and archive data to be made available on-line through ADS. Concerning the data, I urge that every archival set be documented with a good description and a *catalog* of the data. This catalog should be searchable and capable of supporting selection criteria. This will enable the user to select desirable data from the archive.

CASERTANO: As a user of catalog data, I have a confession to make. Every time I have looked for some kind of catalog, I have found it easy to reach within one or the other of the available databases. At least for a researcher in North America or Europe, access to catalog data is only a matter of effort. Yet, I myself have been occasionally guilty of not making that small effort. Users must realize that it is up to them to make use of the facilities that exist. We need to be reminded that databases for catalog data are numerous, easily reachable, and very convenient.

MODERATOR: A discussion of some details specific to spectroscopy follows, I cannot help noting that similar issues can be expected in organizing archives for other kinds of data.

GRIFFIN: In answer to the initial question, "How can we improve access to data and data centers," I would like to report the formation in 1992 of an "IAU Working Group on Spectroscopic Data Archives." This WG aims to stimulate, encourage, and promote the formation of meaningful archives of spectroscopic data and to make them accessible worldwide. The first step is to discuss (in many cases in the face of considerable opposition) the beginnings of banks of raw data, because it is not yet universally accepted that raw data should be preserved. The next step is to construct CONTENTS catalogues for browsing, but ultimately we aim to collect REDUCED spectra as the form of observations which are most readily accessible and usable by other astronomers. We have the responsibility to handle data from the future and the present as well as the past as far as possible, and have a special concern about the presentation of photographic plate archives. In the latter respect, we are currently encouraging the creation of on-time contents catalogues of plate files according to a uniform recipe which we supply.

The profile of data handling, archiving, and accessing will be raised very significantly by educating today's students both in those activities and in the scientific justifications for using data from wide time bases and wide wavelength ranges; that in itself will ensure a developing and lasting conscience toward such activities.

WORLEY: Is your Working Group going to do anything about the archiving of radial velocity data? This seems to be a need which is not being addressed by the astronomical community.

GRIFFIN: No, not as a specific concern. The Geneva Observatory maintains a catalog of measurements made with Swiss Coravels.

MODERATOR: The few astrometric remarks that follow contain a tremendous scientific opportunity; one hopes that the community is sufficiently well-organized to seize it.

ROESER: The observers of the Astrographic Catalogue (AC) about 100 years ago stored the information in the medium of their epoch. They stored the x,y coordinates of their measurements in printed books. Most of the original plates are still available. Do any of the groups presently digitizing sky surveys think of digitizing AC's?

URBAN: A remark concerning AC plate re-measurements: As I remember, about 50% of the observatories that hold AC plates were willing to "give them up" for a time for remeasurement, but few could provide funding.

HUMPHREYS: With regard to plate scanning, it is very likely that in about one decade most of the high-speed measuring machines will be closed. The next generation of surveys will be made with CCDs. Perhaps some effort should be made now to identify large and complete photographic surveys and ensure that they get digitized.

MODERATOR: A major element of the discussion centered on the fact that the data volumes associated with the new archives are quite large.

PIER: How are we going to handle the huge new digital databases just coming into astronomy, e.g., digital scans of POSS, Sloan Digital Survey, large-format arrays of CCDs, etc? For example, at USNO, we do not have the manpower, hardware, or software to support community access to the raw data "scans" (footprints) of the PMM measuring machine.

LASKER: I think all the sky survey groups, photographic or electronic, are concerned with this.

VAN ALTENA: We should also remember that large format CCDs produce an enormous amount of data that is not currently being archived.

ROMAN: In talking about data volume, astronomers do not know what large data sets are. The earth observation satellites will produce gigabytes per day. Millions of dollars are being spent on developing ways to handle these data. If we wait a few years, we may be able to ride on their coat-tails.

RICHMOND: One idea we in the Digital Sky Survey have is to combine *raw* data and *calibration* by extracting information in *real-time* from the database, as it is requested. One can query the database to ask for the current calibration constants and algorithms; they are, in a sense, never separated until the user makes his own copy.

CREZE: When it comes to deal with hundreds of terabytes of data, the real question is not how to store it, but rather how to make scientific use of it. I believe if the latter question has been properly addressed, then the former is likely to be marginal.

RICHMOND: Big projects should make cost-benefit analyses of archiving. If it costs more to create and maintain archives than to acquire information, maybe we should put less emphasis on archiving.

GRIFFIN: The *scientific* justification for creating and maintaining astronomical archives is well known and needs not be repeated here. However, one aspect that is not always emphasized adequately is a *financial* one. Many astronomical activities are, to a greater or smaller degree, financed through public money. If astronomers do not believe in the long-term value of their data and they are thus openly refusing to maximize their use, they may have to face reciprocal attitudes from the public when they expect to continue to finance projects. *Public and conscientious accountability* is therefore of major importance.

MODERATOR'S SUMMARY: That this has been such a spirited and far-ranging discussion leads me to believe that good archiving is something most of this group holds to be of great value. While there can be as many summaries as there were participants, my own view is to direct our summary to the identity of the data, to its user-friendliness, and to its volume. I feel there is a consensus that we should take the term data as broadly as possible, including raw data, the results obtained therefrom, and some measures of their quality and limitations. Equally, a full understanding of the data requires archiving of the details of the instrumentation used and its calibration.

In the general area of user-friendliness, we affirmed the importance of being able to find the wanted data, of having it well-documented and being supported with good search tools, and of having it in a usable form.

Matters related to data volume received considerable attention and are clearly a problem at present. However, given expectations of technical advances, including those driven by non-astronomical databases, the expectations are optimistic.

dy # PHOTOGRAPHIC SURVEYS AND THEIR DIGITIZATION

Barry M. Lasker

Space Telescope Science Institute

1. THE PHOTOGRAPHIC SURVEYS

Historically, three large Schmidt-type telescopes have been dedicated to formal surveys characterized by all-sky coverage and widespread community distribution of the images. These, of course, are the Oschin Telescope (Minkowski and Abell 1963; Reid, et al. 1991), the UK Schmidt Telescope (Cannon 1984), and the ESO Schmidt Telescope (West 1984). While a motivation for the support of these telescopes and their surveys was generally their application as "finder" instruments for associated large telescopes (see, e.g., Minkowski 1972), the scientific impacts of the Schmidt surveys have been far broader. A discussion of the technology and science of Schmidt surveys is well beyond the scope of this paper, and the reader is referred to the major symposia on the subject, i.e., at Hamburg (Haug 1972), Asiago (Capaccioli 1984), Geneva (Jaschek 1989), Edinburgh (MacGillivray and Thompson 1992), and Potsdam (West, et al. 1994).

A few additional comments seem fitting. The very strong institutional commitment to the major surveys is the foundation of their success and scientific significance. This commitment underlies the very high level of quality control we have come to expect, both in the exposure and in the copying, as well as the support of general distribution programs. Until recently, distribution has been on photographic media; but the publication of digitized data products, both images and catalogs derived from them, is now very important.

Data for most major surveys have been tabulated by Morgan et al. (1992), to which we provide some addenda on the northern surveys. A circa 1977 red survey to the sky limit on 103aE emulsion was made at Palomar to search for large proper-motion objects (Luyten 1987, Humphreys 1993); this is archived at the University of Minnesota. A circa 1983 yellow survey to approximately 19 mag on IIaD emulsion was made at Palomar to furnish recent-epoch positions in support of HST operations (Lasker, et al. 1990); this is archived at the Space Telescope Science Institute. Neither of these surveys is likely to be distributed photographically. Finally, we note that all POSS-II blue exposures were made with a GG385 filter (not the GG395 sometimes cited; Reid, private communication).

Each of the survey telescopes has taken a huge number of plates that are not part of any formal survey. The older of these plates are especially valuable, e.g., for proper-motion baselines and for studies of variability; and it is conceivable that appropriate arrangements can be made to incorporate these materials into new programs oriented around fast measuring machines. Another inadequately explored opportunity lies with the near infra-red surveys (IV-N plus RG9 [north] or RG175 [south]). These can provide valuable cross-matching and checking for the

Workshop on Databases for Galactic Structure
A. G. Davis Philip, Bernard Hauck and Arthur
R. Upgren, eds. p. 77 - 86 © 1993 L. Davis Press

Barry M. Lasker
Space Telescope Science Institute
3700 St. Martin Drive
Baltimore, Maryland
21218

(electronic) surveys soon to be initiated in the one micron region.

Finally we should note that new industrial developments are likely to change our approaches to sky surveys. In the purely photographic domain, there may well be changes in the availability of emulsions (Brown 1993). It is possible that film may become a preferred medium at the telescope (Russell et al. 1992); and if film availability continues to be limited to "panchromatic" spectral sensitivities, we may need to discover new filter configurations to implement the traditional blue and yellow passbands (recall that most popular blue/yellow filters have very strong red transmission). The other exciting technical development lies in the maturation of electronic detectors and the number of planned or proposed digital sky surveys based on them. Their intrinsic sensitivity and linearity promise a great advance in the quality of the calibrations and the depth of coverage. In this reviewer's opinion, parity between the photographic and the electronic surveys (in actual catalog production, as opposed to technical capability) will occur around the turn of the century.

2. PLATE COPYING.

The perennial debate about the relative merits of working with original plates as opposed to film or glass copies is likely to continue as long as we measure photographic materials. For a number of important surveys which are very unlikely candidates for photographic copying, it is important that the capability to measure originals exists.

However, for the widely copied and distributed surveys, it is appropriate to note a few original versus copy issues. Perhaps foremost from the viewpoint of practical microdensitometry is the range of densities. Original plates are exposed to a sky density that gives the best threshold sensitivity; typical sky values are 1.5 to 2.0 for the newer surveys, 1.0 to 1.5 for the older; and the saturation densities are in the range 3 to 5. For classical microdensitometers illuminated by (thermal) white light sources, this presents a photon budget problem that severely limits the pixel times, e.g., to hundreds of microseconds - far too slow for high speed scanning; however, this does not appear to be a fundamental (as opposed to engineering) issue for multi-element detectors and laser flying spot systems.

Copying is generally optimized for fidelity with threshold images near the plate limit, and the inevitable result is a loss of information at higher densities. Clearly the significance of this varies according to the scientific goal (e.g., King's 1978 comment on galaxy classification).

Comments on the copying process and on its photometric and astrometric fidelity may be found, e.g., in Standen and Tritton (1979), West (1978) and van Haarlem et al. (1991).

These investigations may be taken as indicating that global copying infidelities exist at the level of a few hundredths of a density unit photometrically and a few microns positionally. The seriousness of these errors depends on the intended methods of calibration. For full-plate reductions based on a few standard stars at the center or on the traditional plate-model approach to astrometry, we may expect the field effects to dominate over the copying errors. However, experience has shown that local photometric solutions to 0.05 mag are possible and that local astrometric solutions constrained only by statistics imposed by the reference catalogs (e.g., Taff, Lattanzi and Bucciarelli 1990; Lattanzi and Bucciarelli 1991) are achievable.

Following the discussion of Lee and van Altena (1983), we also note that if the local accuracy is limited only by the properties of the photographic material of the original plate, one can approach 0.3 microns in position and 0.01 mag in intensity. It is this reviewer's opinion that the definitive study of the local properties of copy material at these levels remains to be done.

3. SELECTED TOPICS IN MICRODENSITOMETRY

Systematic digitization of the Schmidt surveys is done primarily at the five sites; these are listed, together with selected characteristics of the equipment and programs, in Table 1. Other excellent digitization facilities (c.f., presentations in MacGillivray and Thompson [1992] and in the newsletters of the IAU Working Group on Wide-Field Imaging) are not addressed in this review, as their activities are oriented more to individual programs than to the systematic processing of entire surveys.

A detailed comparison of the performance of the machines is best deferred because (a) some of them are mature devices while others are still in development and (b) the more interesting discussion, i.e., about their astrophysical products, must await the completion and distribution of more of the programs. Rather, at this time it is more useful to focus attention on a few astrometric, photometric, and archiving issues common to all digitization programs.

3.1 Metrology Testing

Traditionally, the astrometric performance of a photographic measuring system has been based on direct-reverse (D-R) tests; the significance of these lies in the pragmatic credibility that comes from comparing two separate sets of measurements of the same images of real objects. (Engineering-type tests, e.g., measurements of a standard ruling, are also useful, although they alone are not a predictor of astrometric performance.) The D-R techniques were first developed and refined within the context of the astrograph measurements. For those materials, good models of the full-plate solution are known to exist; and so it is natural that the D-R tests be done on full-plate data. However, for Schmidt plates it is now known that the astrometry is amenable to simple models only over relatively small areas. Lattanzi and Bucciarelli (1991) give a typical covariance length of one-half degree; and Taff, Lattanzi and Bucciarelli (1990) show that (given an adequate grid of reference stars), reductions may well be done with an overlapping grid of subplates with size, again, of order one-half degree.

It is therefore reasonable to consider the characteristics of D-R tests for plates reduced in small pieces, i.e., with areas comparable to subplates. Assuming the eventual availability of excellent reference catalogs of adequate density ($\sigma < 0\rlap{.}''1$, N $\sim 200/$sq deg, e.g., deVegt 1989), we may expect the definition of the reference frame to be at the 0.2 μ level, which is in fact less than the 0.3 μ level to be expected from type III emulsions. The goal for the measuring process should be to avoid compromising this reference frame.

The only D-R data readily available to this reviewer are from the commissioning tests on GAMMA machine (see Sturch et al. 1993, this volume). For this analysis, the full plate was divided into a grid of 13^2 cells, each of about 0.5 degree. A contour plot of the cell rms residuals about the cell mean is given in Fig. 1 for the y-measurements; the x plot is similar. The error for a single datum is assumed to be the D-R local errors divided by $2^{1/2}$, i.e., about

TABLE 1

MEASURING MACHINES DEDICATED TO ALL-SKY SCHMIDT SURVEYS

Name	Principle of operation	plate time (hr)	pixel size (μ)	data products	Refs	Site
COSMOS[1]	CRT flying spot	6	16	c,t	a, c	
SuperCOSMOS[2]	2048 element linear array	2	10	c,t	b, c	ROE
APM	Laser beam with AO deflector	5	7.5	c,t	d, e	Camb.
GSSS PDS	Classical microdensitometer, modified	30	15[3]	c,s	f	
GAMMA[2]	Laser beam with AO deflector (on PDS substrate[4])	10	15	c,s		ST ScI
APS	Laser beam with spinning deflector	2	5 x 12	c,t	h, i	Minn.
PPM[2]	1320 X 1035 element CCD cameras	3	14	c	j, k	USNO

Almost invariably, the listed machines are easily reconfigurable, and so the indicated speeds and pixel sizes are only indicative of typical configurations. The literature citations are by no means complete; rather an attempt has been made to provide a few references from which others may be found. The data products are coded as follows: c, catalogs; t, thresholded cutouts of detected objects; s, full-plate scans (sometimes compressed).

Notes and References to Table 1:

1. Superseded by SuperCOSMOS in 1993 (MacGillivray 1993).
2. Still in commissioning phase.
3. Considerable early work was done with 25 mu pixels.
4. Original white-light microdensitometer capability retained as an option.

a. MacGillivray and Beard (1989).
b. Miller, et al. (1992).
c. MacGillivray (1993).
d. Kibblewhite, et al. (1984).
e. Irwin and McMahon (1992).
f. Lasker, et al. (1990).
g. Sturch, et al. (1993).
h. Pennington, et al. (1992).
i. Pennington, et al. (1993).
j. Monet and Westerhout (1989).
k. Pier and Monet (1989).

0.53 μ. We refrain from a detailed discussion of the $2^{1/2}$ factor or a decomposition of the result into effects due to centroiding, environment, and metrology errors; those details, while appropriate to detailed presentations for specific machines, are a distraction here. The important point is the reformulation of the D-R procedure in light of the modern approach of doing Schmidt astrometry locally; this testing method merits refinement and more widespread use.

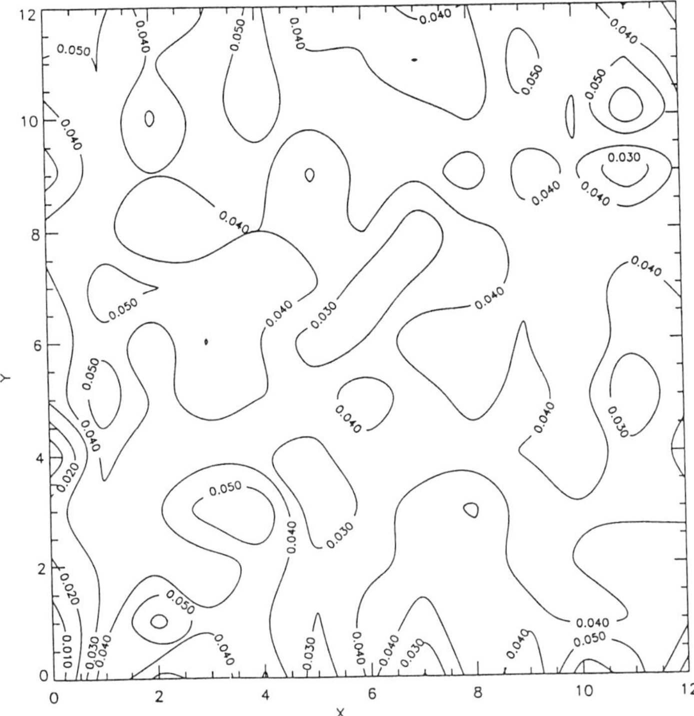

Fig. 1. Contour plots of a direct-reverse test made with the analysis approach described in the text. The y-residuals are shown, and a clipped standard deviation has been used so as to exclude the effects of the occasional bad centroid. The contour intervals are in units of 15 μ pixels, and the abscissa and ordinate delineate the thirteen cells on the plate.

3.2 Diffraction Effects in Laser Systems

Three of the measuring systems cited in Table I use laser illumination (as does a machine used in solar research, Arrambide, et al. 1984). Swing (1976) analyzes the optical requirements for linear photometric response in a microdensitometer. White-light systems generally (at least under easily achievable astronomical scanning conditions) have linear photometric properties; however, laser-illuminated systems, because of diffraction effects associated with the coherence of the wavefront, have a very strong (in practice unachievable) restriction on their linearity, namely that all light in the egress beam must be collected.

By definition, the systems we are considering sample astronomical objects well. The spot size is thus small compared to the *image gradients*, and the dominant diffraction effects are associated with the emulsion *grain*, which has a scale comparable to the laser wavelength. The resulting diffraction pattern is related to the power spectrum of the granularity and (to the extent that other small-scale detail exists) the structure of the image. A prototype system exhibiting full linearity in this domain has been designed and demonstrated (Fallon 1973; Cronin and Reynolds 1973). An essential feature thereof is an egress lens optically contacted (oiled) to the plate so as to collect rays with large diffraction angles, which would otherwise suffer total internal reflection in the glass; however, this is not a practical approach to high-speed astronomical scanning.

It is clear from the reports of the various measuring groups that laser-illuminated systems do give good practical results. Apparently, light loss due to incomplete collection by the egress optics is relatively constant with image conditions and plate density. This is perhaps less surprising when we recall that the plates are typically exposed to sky densities of 1.5 to 2.0, such that we are only concerned with the power spectrum of the emulsion granularity for densities between that level and saturation. One also observes that photometric non-linearities may become invisible because of a compensatory calibration process, e.g., with sensitometer spots or non-linear fitting to reference stars; however, negating the significance of such effects is the fact that the APM (laser-illuminated) accurately reproduces spot measurements from the RGO PDS (Irwin 1993). It would be of some value to put these speculations on a more solid foundation; additionally, the existence of these diffraction issues lends importance to the validation of new laser-illuminated scanning configurations with data obtained from white-light systems.

4. ARCHIVING

A convenient measure for evaluating archival requirements and capabilities is the volume of an all-sky plate set, digitized, in one color, on a 5 degree grid. Such a data set consists of about 1800 plates, with corresponding(uncompressed) sizes ranging from 1.8 Tbyte for 16 μ pixels to 8 Tbyte for 7.5 μ pixels; and a more realistic estimate, allowing for the availability of three passbands (e.g., J, F and N), is three times larger. Such archive sizes are no longer at the frontiers of storage and database technology; however, the proper organization, storage, access, and distribution of such a volume does require considerable care.The storage technologies used to support scanning programs are mainly Exabyte tape, capacity 2-5 Gbyte per volume (all capacities in this section are approximate and must be expected to vary as technology evolves), digital audio tape (1 Gbyte), several kinds of small-format read/write optical disks(0.3-0.5 Gbyte per side), the older large format optical disks (1 Gbyte per side), and the newer ones (2-5 Gbyte per side). Major digitized plate collections accordingly consist of hundreds to thousands of volumes.

The technical characteristics and costs of disks (random access, good long-term stability, typically a few hundred dollars per volume) as opposed to tapes (sequential access only, with long-term stability requiring an ongoing copying/exercising program, typically tens of dollars per volume) needs to be evaluated in the context of each application. Many programs concentrate on a large number of *small areas*, i.e., making finding charts, optical identification of objects discovered at other wavelengths, and operational problem-solving at observatories. Interesting published accounts of this type of work in the context of optical identifications for ROSAT sources may be found in McLean and Burg (1992) and in Yentis (1992). For these, random access is an essential condition for practical work on a large scale; and the appropriate technology is clearly disk-based. On the other hand, catalog construction is based on large-area access, i.e., with data spanning many plates, or even the entire sky. For this kind of work, the sequential properties of tapes are less of a concern.

Judgement as to what to archive varies considerably among the measuring groups. Some groups archive their raw pixels as a resource for both generalized image access and cataloging (ST ScI), or for the contingency of reprocessing their object lists (USNO); others, relying more heavily on their measuring machines, keep only catalog products and thresholded images, with the option of rescanning plates as required by future projects (APM, APS, COSMOS). We note that the storage requirements for the thresholded data is about a factor often lower than for the pixel data.

While storage of the Tbyte archives at the scanning sites is tractable with any combination of the options listed above, their distribution for community use raises other issues. *Copying* the data is costly both in terms of computer time and labor; it can be done only on a small scale for special projects or on a large scale for unique collaborations (e.g., the duplication of ROE data-base at NRL; Yentis et al. 1992). Accordingly, other approaches more appropriate for the practical and economic dissemination of the data bases on a larger scale are currently receiving considerable attention. The CD ROM technology is economical because the disks are produced by *printing* (as opposed to computer-copying) techniques and because the pricing has been driven down by the popularity of the medium; applications are discussed in the papers by Garcia (1993) and by Sturch et al. (1993) (both elsewhere in this volume). Where access to only small areas of the sky is required, *network access* is also an excellent option if adequate resources (jukeboxes or labor) are available at the central site; applications are discussed by Eichhorn (1993) and by Humphreys et al. (1993) (both elsewhere in this volume). However, one must raise the obvious caution, that in certain parts of the world, good network connectivity may not be available for some time.

5. EXPECTATIONS

At the time of this writing, three of the systems in Table 1 were still in their implementation phases; however, it is likely that the coming year will see all five in stable and routine operational configurations. The following five years may well be regarded as the prime of Schmidt photography and high-speed measuring machines. Productivity will be high and important research contributions are to be expected. Around the turn of the century, parity may be expected from the all-electronic programs (e.g., Richmond et al. 1993) and even then photographic surveys will be a major resource for archival proper-motion and variability information. Indubitably, every measuring and cataloging system has its own defects and systematic effects, (colloquially, "warts"), for which the best control is checking results against independently produced data sets. Fortunately, there is enough overlap in the programs of the various groups that these checks are possible. Indeed, the availability of independently produced data sets raises an even larger opportunity: that of a "consolidation catalog", selected from the best material of each group, and including careful reconciliation of the discrepancies. It does not appear that this important project is within the scope or interest of any of the measuring groups, but it is one for which the community would benefit greatly from a well-coordinated collaborative effort. The author thanks M. Damashek, H. Jenkner, V. G. Laidler, M. Lattanzi, B. J. McLean, and W. van Altena for many stimulating conversations on the topic of this review. The ST ScI is operated by the Association of Universities for Research in Astronomy, Inc., under contract to NASA.

REFERENCES

Arrambide, M. R., Dunn, R. B., Healy, A. W., Porter, R., Widner, A. L., November, L. J. and Spence, G. E. 1984 in Astronomical Microdensitometry, D. A. Klinglesmith, ed., NASA CP-2317, p. 277

Brown, G. P. 1993 in Working Group on Wide-Field Imaging, Newsletter No. 3, H. T. MacGillivray, ed., Royal Observatory Edinburgh, p. 88

Cannon, R. D. 1984 in Astronomy with Schmidt-Type Telescopes, M. Capaccioli, ed., Reidel, Dordrecht, p. 25

Capaccioli, M., ed., Astronomy with Schmidt-Type Telescopes Reidel, Dordrecht

Cronin, D. J. and Reynolds, G. O. 1973 Optical Engineering, 12, 201

deVegt, Ch. 1989 in Digitized Optical Sky Surveys, C. Jaschek, ed., Bull. Inform. du CDS, Number 37, p. 21

Fallon, J. P. 1973 Optical Engineering 12, 206

Garcia, M. 1993 in Workshop on Databases in Galactic Structure, A. G. D. Philip, B. Hauck and A. R. Upgren, eds., L. Davis Press, Schenectady, p. 9

Haug, U., ed., 1972 Conference on the Role of Schmidt Telescopes in Astronomy, Hamburger Sternwarte Bergedorf, Hamburg

Humphreys, R. M. 1993, private communication

Irwin, M. 1993, private communication

Irwin, M. and McMahon, R. 1992 in Working Group on Wide-Field Imaging, Newsletter No. 2, H. T. MacGillivray, ed., Royal Observatory Edinburgh, p. 31

Jaschek, C., editor, 1989, Digitized Optical Sky Surveys, in Bull. Inform. du CDS, No. 37

Kibblewhite, E. J., Bridgeland, M. T., Bunclark, P. and Irwin, M. 1984 in Astronomical Microdensitometry, D. A. Klinglesmith, ed., NASA CP-2317, p. 277

King. I. R. 1978 query to presentation of West (1978)

Lasker, B. M., Sturch, C. R., McLean, B. J., Russell, J. L, Jenkner, H. and Shara, M. 1990 AJ 89, 2019

Lattanzi, M. G. and Bucciarelli, B. 1991 A&A 250, 565

Lee, J. -F. and van Altena, W. 1983 AJ 88, 1683

Luyten, W. 1987 Proper Motion Survey with the Forty-Eight in Schmidt Telescope, and previous publications in the same series, Univ. of Minn., Minneapolis

MacGillivray, H. T. 1992 in Working Group on Wide-Field Imaging, Newsletter No. 2, H.T. MacGillivray, ed., Royal Observatory Edinburgh, p. 34

MacGillivray, H. T. and Beard, S. M. 1989 in Digitized Optical Sky Surveys, C. Jaschek, ed., Bull. Inform. du CDS, Number 37, p. 65

MacGillivray, H. T. and Thompson, E. B. 1992 Digitized Optical Sky Surveys, Kluwer Academic Pub., Dordrecht

McLean, B. J. and Burg, R. 1992 in Digitized Optical Sky Surveys, H. T. MacGillivray and

E. B. Thompson, eds., Kluwer Academic Pub., Dordrecht, p. 465

Miller, L., Cormack, W., Paterson, M., Beard, S. and Lawrence, L. 1992 in Digitized Optical Sky Surveys, H. T. MacGillivray and E. B. Thompson, eds., Kluwer Academic Pub., Dordrecht, p. 123

Minkowski, R. L. 1972 in Conference on the Role of Schmidt Telescopes in Astronomy, U. Haug, ed., 1972 Hamburger Sternwarte Bergedorf, Hamburg, p. 5

Minkowski, R. L. and Abell, G. O. 1963 in Stars and Stellar Systems, vol. III., K. Aa. Strand, ed., Univ. of Chicago Press, Chicago, p. 481

Monet, D. and Westerhout, G. 1989 in Digitized Optical Sky Surveys, C. Jaschek, ed., Bull. Inform. du CDS, No. 37, p. 75

Morgan, D. H., Tritton, S. B., Savage, A., Hartley, M. and Cannon, R. D. 1992 in Digitized Optical Sky Surveys, H. T. MacGillivray and E. B. Thompson, eds., Kluwer Academic Pub., Dordrecht, p. 11

Pennington, R. L, Humphreys, R. M., Odewhan, S. C., Zumach, W. and Stockwell, E. B. 1992 in Digitized Optical Sky Surveys, H. T. MacGillivray and E. B. Thompson, eds., Kluwer Academic Pub., Dordrecht, p. 77

Pennington, R. L, Humphreys, R. M., Odewhan, S. C., Zumach, W. and Thurmes, P. M. 1993 PASP 105, 521

Pier, J., and Monet, D. 1993 in Workshop on Databases for Galactic Structure, A. G. D. Philip, B. Hauck and A. R. Upgren, eds., L. Davis Press, Schenectady, p. 161

Reid, I. N, Brewer, C., Brucato., R. J., McKinley, W. R., Maury, A., Mendenhall, D., Mould, J., Mueller, J., Neugebauer, G., Phinney, J., Sargent, W. L. W., Schombert, J. and Thicksten, R. 1991 PASP 103, 661

Russell, K., Malin, D.F, Savage, A., Hartley, M. and Parker, Q.A. in Digitized Optical Sky Surveys, H. T. MacGillivray and E. B. Thompson, eds., Kluwer Academic Pub., Dordrecht, p. 23

Standen, P. R. and Tritton, K. P. 1979 Occ. Reports of ROE, Number 5

Sturch, C. R., Laidler, V. G., Greene, G. R., Lasker, B. M. and Postman, M. 1993 in Workshop on Databases for Galactic Structure, A. G. D. Philip, B. Hauck and A. R. Upgren, eds., L. Davis Press, Schenectady, p. 201

Swing, R. E. 1976 Optical Engineering, 15, 559

Taff, L. G., Lattanzi, M. G. and Bucciarelli, B. 1990 ApJ 358, 359

van Haarlem, M. P., LePoole, R. S., Katgert, P. and Tritton, S. 1992 MNRAS 255, 295

West, R. M. 1978 in Modern Techniques in Astronomical Photography, R. M. West and J. C. Heudier, eds., ESO, Garching, p. 193

West, R. M. 1984 in Astronomy with Schmidt-Type Telescopes, M. Capaccioli, ed., Reidel, Dordrecht, p. 13

West, R. M., et al. 1993, eds., IAU Symposium No. 161, Wide-Field Imaging

Yentis, D. J., Cruddace, R. G., Gursky, H., Stuart, B. V., Wallin, J. F., MacGillivray, H. T., and Collins, C. A. 1992, in Digitized Optical Sky Surveys, H. T. MacGillivray and E. B. Thompson, eds., Kluwer Academic Pub., Dordrecht, p. 67

DISCUSSION

VAN ALTENA: I am not sure if it is possible to produce a composite catalogue since each catalogue is designed for a specific purpose, e.g., astrometry or photometry. At Yale, I. Platais has made a combination of the ROE Object Catalogue and the HST Guide Star Catalogue that is used as an input catalogue for our PDS microdensitometer. No doubt other groups would find another combination that suited them better.

LASKER: A composite catalog is most certainly a difficult undertaking that calls for many difficult scientific judgements along the lines you indicate. However, the community would benefit greatly from the best collective judgement of the scanning community as to what the sky is like at fainter magnitudes.

CATALOGS AND DATABASES FROM PHOTOGRAPHIC AND CCD SKY SURVEYS AND THEIR APPLICATIONS TO GALACTIC STRUCTURE

Roberta M. Humphreys

University of Minnesota

1. INTRODUCTORY REMARKS

Several years ago I began to think about what kinds of science would be possible if we had the "whole sky" of stars and galaxies available on-line, that is, accessible via our computers.

With the Automated Plate Scanner and a set of glass copies of the original Palomar Sky Survey at Minnesota, the next steps were obvious - digitizing the plates, producing the object catalog and a database for access and display of the data. Several other measuring machine groups have similar programs in progress, and in the previous paper, Barry Lasker (1993) has very thoroughly described the different digitization surveys using photographic plates.

However, the role of the photographic plate in astronomy is definitely on the decline. The second epoch Palomar Sky Survey will be the last major photographic survey in this country. The UK Schmidt may continue with photographic surveys for another decade, but beyond that any further photographic surveys are increasingly doubtful. Therefore, it is no surprise that major sky surveys with CCD arrays are now planned.

2. THE DIGITAL SKY SURVEYS

I will just briefly summarize some of the digital sky surveys. This is not intended to be a comprehensive list, but to give an overview of what can be done with the digital surveys, most of which are still in the planning stage.

2.1 Optical

1. The CCD Transit Instrument (CTI) at the University of Arizona under the direction of John McGraw is the first of the large digital surveys. As one of the first, CTI experimented with and developed many of the techniques now adopted by the other surveys such as use of the TDI (scanning) mode with a CCD. The CTI Survey is a limited strip survey at $+28°$ with a 71-inch telescope on Kitt Peak. Observations are made in the V filter every night and also in B, R or I with the second CCD. With repeated scans of the same region of the sky, the data are very useful for searching for transient events and variability.

2. The Sloan Digital Sky Survey (Princeton, Chicago, Institute for Advanced Study, FermiLab, Johns Hopkins and Japan) is a CCD imaging and spectroscopic survey of the North Galactic Cap and a more limited survey in part of the southern Galactic hemisphere. The

Workshop on Databases for Galactic Structure
A. G. Davis Philip, Bernard Hauck and Arthur
R. Upgren, eds. p. 87 - 95 © 1993 L. Davis Press

Roberta M. Humphreys
Dept. of Astronomy
University of Minnesota
Minneapolis, Minnesota
55455

survey will be done with a 2.5-meter telescope and large CCD array camera and two large format fiber-fed spectrographs. The imaging data will be obtained in five wavelength (u, g, r, i, z) bands with an expected limiting magnitude of 23.

The emphasis of the survey is on cosmology and large scale structure, although a large amount of multicolor data will also be available for stars. The spectroscopy is for galaxy and quasar redshifts. The observations are expected to begin in 1995 and be completed about the year 2000. A more complete description is given in the paper by Michael Richmond (1993) in this volume.

3. The Large Imaging Telescope (LITE), a European (French/German) project is a CCD survey of the southern sky similar to the Sloan Survey. It will also use a dedicated 2.5-meter telescope; the survey will be multi-color and they plan to go fainter than Sloan. Like SDSS the focus is mainly cosmological. Follow-up spectroscopy will be done with the VLT. Starting date is estimated for 1999.

4. The Spaceguard Survey is a proposed network of telescopes equipped with CCD cameras to search for near-Earth objects. The expected limiting magnitude would be 22 mag. A survey of this type would serve not only as a search for asteroids and comets, but also for other transient events. The status of this proposal is uncertain.

Note that except for the CTI project, all of these large digital surveys are still in the planning stage and until completion of their observations and the production of their catalogs, the plate digitization surveys will provide most of our information about large areas of the sky in the optical region.

2.2 Infrared

The original Two Micron Sky Survey by Neugebauer and Leighton (1965) has a magnitude limit of only 3.5 mag, while the 12 μ limit of IRAS is about 4000 times fainter in flux. (The optical surveys of course go much fainter.) Thus there is clearly a large gap in the flux limit of our sky coverage between 1 and 5 microns. The 2MASS project, under the leadership of Susan Kleinman at the University of Massachusetts, is a proposed all-sky infrared survey at J, H and K' (2.16μm) that will fill this gap in our wavelength coverage of the sky. The dedicated 1.5 meter telescopes, one in each hemisphere, will be equipped with three 256 x 256 HgCd Te arrays. The expected limit at K' is 14 mag at the 10σ level. DENIS is a comparable near-infrared survey of the southern sky by a consortium of European institutions. The observations will be made at I, J and K' with 1σ limits of 18, 16 and 14 magnitudes, respectively, with a dedicated 1-meter telescope on La Silla. This project is expected to start in 1994. DENIS is described in the paper by Epchtein (1993) in this volume.

3. THEIR CATALOGS AND DATABASES

In this summary I will distinguish between a catalog of objects and their associated parameters and a database or Database Management System (DBMS). An ideal DBMS is designed for rapid access, display and manipulation of vast amounts of data.

3.1 Existing Catalogs and Databases

Guide Star Catalog - the first release (Lasker et al. 1990, Russell et al. 1990 and Jenkner et al. 1990) is available on CD-ROM. It contains data for 2 x 10^7 stars including positions and magnitudes for the magnitude range V = 9 to 16th Mag.

CTI - a database derived from the CTI survey. It includes multiple entries for all detections - positions, magnitudes in four wavelengths and object classification.

COSMOS/UKST - Catalog of Southern Sky, produced by an ROE/NRL collaboration, is an object catalog from the Cosmos scans of UK Schmidt plates IIIaJ and the short red plates south of +2.5 degrees. It includes more than 50 x 10^7 objects down to the plate limit. Requests to use the Catalog can be made to ROE or NRL.

The APM group at Cambridge, England also has scans of the UK Schmidt plate material, and they have produced a catalog from their scans of the POSS I O and E plates. The catalog includes more than 10 x 10^7 objects with b > ¦30°¦. Access is by request for a limited subset. Additional data processing in Cambridge or access to the entire dataset is available as a collaboration.

3.2 In Preparation

APS Catalog of the POSS I - The results of cataloging the APS scans of the POSS I blue and red plates for b > ¦20°¦. The DBMS - StarBase permits rapid access and display of the data. The Catalog and StarBase may be accessed on-line directly across Internet and via NASA's Astrophysical Data System. The APS will be a node on ADS. The characteristics of the APS Catalog and StarBase are described by Pennington et al. (1993) and in the paper by Humphreys et al. (1993) in this volume. The database will have approximately 400 Gigabytes including the Catalog and pixel data for the non-stellar images.

STScI - The catalog of scans of the POSS I E plates and the SERC J plates to be available on 100 CD-ROM's this year.

3.3 Planned and Future Databases from the Large Digitized and Digital Surveys

APS - scans of the "Luyten-red" plates, a second-epoch Palomar Sky Survey for proper motions and faint red stars. POSS II

STScI - POSS II

USNO - POSS I and II for astrometry and proper motions.

APM and Cosmos - additional colors in southern Schmidt Surveys [UKST-J, ESO-B, ESO-R, AAO-R (2nd epoch)] etc. POSS II

Sloan Digital Sky Survey - 6.5 Terabytes of image data are expected plus 10 Gigabytes for the spectra. Plans for publication include an atlas of detected objects plus various specialized catalogs including the spectra.

4. GALACTIC STRUCTURE FROM VERY LARGE DATABASES

A very large database of several 100 million stars complete to some faint magnitude will obviously have numerous applications to statistical studies where completeness and uniformity are important. Counting stars to determine the properties and distribution of different populations is a well established method (see Bok 1937). Modern applications of this method are the pioneering studies by Bahcall and Soneira (see review by Bahcall 1986) and by Gilmore and Reid (see review by Gilmore et al. 1989) in which mathematical models of the galaxy are compared with the observed counts.

Infrared observations have also been used to model the galaxy using the 2μ data and the IRAS point-source data (for example see papers by Jones et al. 1981, Habing et al. 1985, Garwood and Jones 1987, Wainscoat et al. 1992 and Cohen 1993). Galactic models from infrared data are discussed in the paper by Martin Cohen (1993) in this volume. Some of the problems with the infrared model include the lack of a good local luminosity function in the infrared and the limits of the IR star counts which do not go faint enough to distinguish halo and disk stars at the poles.

In most cases the optical studies have relied on star counts from different sources with different filters, magnitude limits and covering areas of different sizes on the sky. This is very clear from Fig. 1 in Bahcall (1986). Obviously a uniform data set could reduce the uncertainty about completeness and small number statistics with star counts to faint limiting magnitudes over a large region of the sky.

One of the very serious limitations of many previous star count studies was the lack of accurate and effective star/galaxy separation. It was either done by visual inspection or more automated methods, both of which were subject to significant errors especially near the limit of the plates. Uniform, accurate and efficient object classification is essential for the counts of both stars and galaxies and should be possible with the new digitized and digital surveys. At Minnesota we have experimented with an automated neural network classifier and have achieved success rates of 95% down to 20th magnitude on both the blue and red POSS plates (Odewahn et al. 1992, 1993) and we are optimistic about extending it to close to the plate limit.

Star counts in only one wavelength can be used to study the luminosity function and density distribution, but identification and separation of different stellar populations requires color information. This is the benefit of these new surveys: POSS I has O-E; POSS II has J-F and F-I colors, and the SloanDDS with five wavelengths will have four colors including an ultraviolet wavelength which will be critical for measuring extinction and separating low metallicity populations.

Table 1 gives an incomplete list of some of the galactic structure problems that can be studied by star counts from these large databases.

TABLE 1

Galactic Structure Projects from Star Count Data

1. Separation of different stellar populations by color.

2. Disk
        luminosity function for old disk main sequence
        vertical structure of disk
        any extended component - "the thick disk"?
3. Halo
        luminosity function for the halo population
        identification of sub-dwarf population (Halo-main-sequence)
        horizontal-branch stars from blue colors
        shape and extent of halo - axial ratio of halo, is it triaxial?

Like the databases from optical surveys, the 2MASS survey will greatly increase the value of IR star count data for galactic structure studies. One of the obvious applications will be to study the bulge population and to search for a stellar population in the bar. The two micron data in combination with optical data will be used to map extinction in the plane, identify the faint M dwarf population out of the plane (the M dwarfs should dominate at the faint end of the 2 micron surveys), and of course immediate comparison with the faint optical surveys will allow us to identify those peculiar objects not found in the visible.

It is also clear from the many digitization and digital surveys that are planned, that proper motions will be one of the obvious results of the comparisons between POSS I and POSS II, and in the case of the APS project, we will have three epochs for motions - POSS I, "Luyten red", and POSS II.

Most people are familiar with the extensive work and the resulting catalogs of proper motions by Willem Luyten (1976, 1980) from his scanning of the POSS and his second epoch red plates with the APS of old (the CDC/Luyten machine). But the LHS Catalog (Luyten 1976) only included stars with motions greater than $0"5/yr$ and the NLTT Catalog (Luyten 1980) had a minimum motion of $0"18/yr$. The still unpublished LP catalog includes all of the motions $\geq 0"08/yr$, but another limitation of the Luyten catalogs was the apparent requirement that at least many stars had to appear on both the blue and red plates of POSS I.

I will use the APS as an example of the kind of proper motion data we can obtain from the current digitization program. The positional repeatability of the APS is 1 micron (the encoders read to 0.366 micron) corresponding to $0"07$ on the POSS which with a typical 15 year baseline for the "Luyten red" means we have the potential to measure motions to $0\rlap{.}{''}005/yr$ (1σ). For the POSS II with an average 35 year baseline, motions can be measured to $0\rlap{.}{''}002/yr$ (1σ). Thus "believable" motions would be $\geq 0\rlap{.}{''}015/yr$ and $0\rlap{.}{''}006/yr$ from the two surveys. Similar results are expected with the USNO machine and other measuring machines. The USNO astrometry and proper motion program is described in the paper by Jeff Pier (1993) in this volume.

Thus scanning the "Luyten red" and the POSS II will produce motions for millions of stars, and although for those near the measurement limits, the accuracy and quality of the proper motions will not be high, they can be used in statistical studies. When combined with the star count data resulting from these various surveys, which will be complete to 20th magnitude or fainter, the motions will allow a combined study of the structure and kinematics of the Galaxy.

Some of the projects that can be done with the addition of motions to the databases are summarized in Table 2.

TABLE 2

Astronomical Projects for Databases with Proper Motions

1. Search for white dwarfs
2. Search for faint red stars at the end of the main sequence.
3. High proper motion objects in the solar neighborhood.
4. Statistical parallaxes of different populations
5. Kinematics of thick disk (?)
6. Kinematics of the halo population: the halo luminosity function

Once we have complete samples of stars down to some faint magnitude with proper motions for many of them, we can then separate the populations and select stars by a variety of criteria - magnitude, color and proper motion - for future study. Specifically, I have in mind multi-object spectroscopy for radial velocities of hundreds of stars with fiber spectrographs. The spectra will also provide additional information about the stars including temperature, luminosity and composition estimates. With this kind of dataset available for selected regions and directions, we will have the potential to study the dynamical history of the Galaxy.

REFERENCES

Bahcall, J. N. 1986 Ann. Rev. Astron. Astrophys. 24, 577

Bok, B. J. 1937 The Distribution of the Stars in Space, Harvard

Cohen, M. 1993 AJ 105, 1860

Cohen, M. 1993 in Workshop on Databases for Galactic Structure, A. G. D. Philip, B. Hauck and A. R. Upgren, eds, L. Davis Press, Schenectady, p. 131

Epchtein, N. 1993 in Workshop on Databases for Galactic Structure, A. G. D. Philip, B. Hauck and A. R. Upgren, eds., L. Davis Press, Schenectady, p. 97

Garwood, R. and Jones, T. J. 1987 PASP 99, 453

Gilmore, G., Wyse, R. F. G. and Kuijken, K. 1989 Ann. Rev. Astron. Astrophys. 27, 555.

Habing, H. J., Olnon, F. M., Chester, T. J., Gillett, F. C., Rowan-Robinson, M. and Neugebauer, G. 1985 A&A 152, L1

Jenkner, H., Lasker, B. M., Sturch, C. R., McLean, B. J., Shara, M. M. and Russell, J. L. 1990 AJ 99, 2082

Jones, T. J. Ashley, M., Hyland, A. R. and Ruelas-Mayorga, A. 1981 MNRAS 197, 413.

Lasker, B. 1993 in Workshop on Databases for Glactic Structure, A. G. D. Philip, B. Hauck and A. R. Upgren, eds., L. Davis Press, Schenectady, p. 77

Lasker, B. M., Sturch, C. R., McLean, B. J., Russell, J., Jenkner, H. and Shara, M. M. 1990 AJ 99, 2019

Luyten, W. J. 1976 LHS Catalog, Univ. of Minn., Minneapolis

Luyten, W. J. 1980 NLTT Catalog, Univ. of Minn., Minneapolis

Neugebauer, G. and Leighton, R. B. 1965 Two Micron Survey

Odewahn, S.C., Humphreys, R.M., Aldering, G.S. and Thurmes, P. 1993 PASP, in press.

Odewahn, S.C., Stockwell, E.B., Pennington, R.L., Humphreys, R.M. and Zumach, W.A. 1992 AJ 103, 318

Pennington, R. L., Humphreys, R. M., Odewahn, S. C., Zumach, W. and Thurmes, P. 1993 PASP 105, 521

Pier, J. and Monet, D. 1933 in Workshop on Databases for Galactic Structure, A. G. D. Philip, B. Hauck and A. R. Upgren, eds., L. Davis Press, Schenectady, p. 161

Richmond, M. 1993 in Workshop on Databases for Galactic Structure, A. G. D. Philip, B. Hauck and A. R. Upgren, eds., L. Davis Press, Schenectady, p. 207

Russell, J. L., Lasker, B. M., McLean, B. J., Sturch, C. R. and Jenkner, H. 1990 AJ 99, 2059

Wainscoat, R. J., Cohen, M., Volk, R., Wolkner, H. J. and Schwartz, D. E. 1992 ApJS 83, 111

DISCUSSION

LU: I would like to point out that there is a survey for QSO's called BAC (Beijing - Arizona Color Survey) for 500 deg^2 of northern sky with the 0.6-m f/3 Schmidt telescope + 20482 Ford Aerospace CCD of Beijing Astr. Obs. The spectral energy distribution of all objects to V = 21 in the 0.9 deg^2 field will be measured from 3200 to 9000 Å with 15 color-bands. Of the 500 fields, 150 are chosen centered on QSOs, 150 are centered on spiral galaxies, 50 are centered on on spiral galaxies, 150 fields will be randomly selected and 50 fields will be selected for calibration purposes. The survey is specifically designed to (a) find QSOs of all kinds with a range 1.7 < z < 6.0, (b) study the large stable distribution of QSOs as a function of z, (c) determine the interrelationship among the structure, stellar populations, and interstellar media of nearby spiral galaxies, and (d) study the spatial distribution and spectral evolution of galaxies.

The program plans to begin observations in September of 1993 and will involve collaboration between Beijing Astr. Obs., the University of Arizona, Arizona State University, Western CT State, and the Institute of Astronomy of the National Central University in Taiwan. The project plans to be completed within 8 years, a mini survey of 1/4 of the fields within 2 years.

MERMILLIOD: Your database offers interesting possibilities for studying galactic open clusters. Do you intend to do something in this domain?

HUMPHREYS: We have no specific plans to search for star clusters.

LASKER: I wish to call attention to the matter of the photometric sequences needed to support the programs described in this paper, as well as many others presented at this workshop. We know that spatially dependent systematic shifts exist in the photometric sensitivity, with scales of only a few centimeters. There are several efforts, our own included (see paper by Postman et al. in the Edinburgh symposium) to develop an all-sky set of faint photometric sequences in the same grid as the Schmidt surveys. These programs need to progress effectively if we are to make good use of the opportunities presented by the Schmidt surveys, and this is best done by bringing in new participants in the photometric collaborations.

HUMPHREYS: We have several POSS fields for which we have measured photoelectric and CCD photometry in grids across the plates. This is being used to determine the vignetting function as well as the extent of other variations with position on the plate. Our photometry is based on BVR CCD and photoelectric photometry for about 300 POSS fields. We find essentially the same shape for the mag-diameter relation with small zero-point shifts.

ROMAN: As someone who used to worry about scarce funds, I wonder what the justification is for scanning POSS II four times? How much interaction and collaboration is there among the groups to insure complementarity rather than duplication?

HUMPHREYS: APS, STScI and USNO are all scanning POSS II but the machines are different and the emphasis of the three groups is different. The USNO machine and project is designed for positional accuracy. At the APS we have put the emphasis on photometry and accurate star/galaxy separation for which we have developed a new approach using several networks. At STScI they are scanning with a PDS and record every pixel on the plate. At APS and USNO we do not record or save every pixel.

LASKER: The three groups to which the discussion refers are really doing significantly different things, to which the POSS II is but one common element. I feel we should look not at the nominal programatics but what the groups are actually doing and what data products they are distributing. Then its easier to see the differences.

LATTANZI: Proper motions derived from Schmidt plates might be as precise as 2 mas/yr, but a word of caution goes to controlling accuracy to that same level. Very high proper motion stars might be hard to find or you need a very good plate matching algorithm.

HUMPHREYS: The numbers I gave are for what the APS is capable and not what the actual errors will be. We create a parallel database of unmatched regions that can be searched for possible high-motion objects.

GUO: How good photometry can you expect? How do you automate the star/galaxy separation?

HUMPHREYS: The calibrations (magnitude-diameter relations) have a typical rms of 0.1 - 0.15 mag. We are using a neural network approach to object classifications described in two papers by Odawahn et al. (1993, 1993).

DENIS: A DEEP NEAR INFRARED SOUTHERN SKY SURVEY: A NEW TOOL TO PROBE HIDDEN STELLAR POPULATIONS

Nicolas Epchtein

Observatoire de Paris

ABSTRACT: DENIS is the first attempt to carry out an all southern sky digital survey in the near-infrared range. It will be undertaken with a dedicated three channel camera in the I, J and K' photometric bands installed at the 1-meter telescope of the European Southern Observatory in Chile. Expected limiting magnitudes are 18, 16 and 14 in the I, J, K' bands, respectively and the spatial resolution will be 3" in J and K' and 1" in I. Several areas of astrophysics will take advantage of this survey. In our Galaxy, the exploration of hidden stellar populations, the improvement of the Initial Mass Function towards low mass stars, statistics on late-type stars, etc. will undoubtedly lead to an improved description of galactic structure. DENIS is expected to start early in '94 and to last for approximately four years. Data will be released in stages and will be made available to the community as soon as appropriate processing and calibrations will be achieved.

1. INTRODUCTION

Investigations of galactic structure and of stellar population distributions are mainly based on stellar counts and large scale surveys of the sky. The spectral sensitivity limitation of the photographic plates has restricted source counts to the optical range, up to one micron. Our knowledge of the star distribution in our Galaxy and of the structure they trace are, thus, strongly biased towards objects that radiate the bulk of their energy below this limit.

The advent of IR astronomy in the sixties showed that the sky is also populated with "infrared stars" that mostly radiate above one micron, and its appearance is quite different from the "usual" optical sky. These stars are either of low effective temperature or they are surrounded by a circumstellar dust envelope (in accretion or expelled), or deeply embedded in clouds of dust and molecules. These populations are of basic interest since they correspond to both ends of stellar evolution and, precisely for lack of knowledge about their statistical properties, they have been essentially ignored in the "classical theories" of stellar evolution.

Thirty years after the first significant results in infrared astronomy, there is still no complete and comprehensive overview of the sky in the spectral range that prolongs the visible beyond the red (1-5 micron) that would allow serious statistical investigations.

The only presently available document describing the sky at infrared wavelengths is the IRAS survey database. As it covers only the 10-100 micron range, it is biased towards extreme objects with optically thick and cold envelopes. COBE/DIRBE has a much too low spatial resolution to single out individual stars and the TMSS (Neugebauer and Leighton 1969) is far

Workshop on Databases for Galactic Structure
A. G. Davis Philip, Bernard Hauck and Arthur
R. Upgren, eds p 97 - 101 © 1993 L. Davis Press

Nicholas Epchtein
Observatory of Paris
DESPA
F-92195 Meudon Cedex
France

too insensitive to explore the Galaxy beyond approximately 1 kpc from the Sun. It is now essential to link the optical and IRAS ranges with a deep and complete survey. Moreover, a good knowledge of the star and galaxy populations in the 1-3 micron domain, is strongly needed while space observatories, such as the ISO and SIRTF missions, partly aimed at observing targets in this range, are underway. Adaptive optics, that will equip large and very large telescopes of the future, also badly need deep and complete two micron star catalogs to pick up reference objects aimed at wavefront analysis. The improvements of the array detector size and sensitivity have been so spectacular in the '80s that undertaking a full sensitive mapping of the sky, at least up to 2.5 micron, can be contemplated at very low cost from the ground even with an existing telescope. DENIS is the first funded project aimed at achieving this goal in the southern sky. The 2MASS project (Kleinmann 1992) chases similar objectives in the all the sky.

2. SCIENTIFIC OBJECTIVES OF DENIS

The aim of DENIS is to bring about a complete mapping of the southern sky in three photometric bands (IJK'), simultaneously, with a spatial resolution of 3" in the JK' bands and 1" in the I band. The upper limit of the spectral range around 2.4 micron is imposed by the degradation of the array performances due to the prohibitive level of thermal background radiated by an uncooled telescope.

The addition of a visible red band (I) has been preferred to the insertion of the H band between J and K', as proposed for the 2MASS project. There are two main advantages for this choice: the I maps will provide, i) a better link with the optical Schmidt surveys, ii) a spatial resolution three times better than in the JK' bands that will greatly facilitate the separation between stars and galaxies, a major concern of these surveys (Epchtein and Mamon 1992). The final goal of DENIS is to build catalogs and databases of point-like sources and small extension sources and to open these data base to the community through computer networks.

A digital sky survey performed for the first time in a till now uncovered range of wavelengths, will undoubtedly lead to a major breakthrough in several areas of astronomy. Basically, star and galaxy statistics will greatly benefit of the huge databases that will come out of DENIS. It is expected that several tens of millions of stars and several hundreds of thousands of galaxies will be detected. The two micron band is extremely well suited to probing the stellar content of evolved galaxies, and primarily of our Galaxy, simply because the most frequent stars in those galaxies are K and M dwarfs that radiate mainly in the near infrared. The other great advantage of observing in the near infrared with respect to the visible is the much better transparency of the interstellar medium that will allow probing stars in highly obscured regions and young stellar objects embedded in their parental clouds, a property that has been often used to explore limited areas of individual clouds, but never globally.

Giants and supergiants are often surrounded by dust and gas; the dust particles reemit the energy in the near infrared. The IRAS mission has discovered tens of thousands of these objects, however, these data are rarely sufficient to fully characterize the nature of the IRAS sources. Photometric data at about 2 micron enable us to make such characterizations. It is likely that the combination of IRAS and DENIS will give us systematic, statistically complete information on the various phenomena occurring during the mass loss episodes. As an example, it allows a good break up of the carbon and oxygen rich giant stars (Epchtein et al. 1990, Guglielmo et al. 1993). The sensitivity of DENIS will even allow a census of the AGB stars in the Magellanic Clouds, including those with thick dust envelopes. The knowledge of the distance will undoubtedly improve the estimation of the luminosity of these objects.

DENIS is expected to detect all red giants without circumstellar envelopes up to the distance of the galactic center. Thus, many stars will be detected that were up till now hidden by interstellar extinction, those pertaining, in particular, to the inner bulge. The IRAS survey has shown the importance of infrared observations (e.g., Habing 1988) but a major difficulty with the IRAS data is the confusion in the central region of the Galaxy. DENIS should be much less affected by confusion, due to a much better spatial resolution and a spectral range that should be more selective towards stars. Stars in the four kpc molecular ring will be detected. It is a major structural feature of the Galaxy, and few facts are known about its age and origin. The radial and vertical scale lengths of the various stellar populations will be derived.

The COBE/DIRBE mission has shown prominently the internal galactic bulge of stars at short wavelengths. This will be revealed more clearly with DENIS by resolving this bulge into stars. Late-type stars will be detected at great distances from the galactic center; it will be possible to trace, for the first time, the stellar distribution in the regions characterized by the galactic warp and dominated gravitationally by dark matter. Substantial progress is thus expected in the description of the structure, evolution and dynamics of the Galaxy. Other galactic structure features should be revealed. Recently formed stars are prominent at two microns and are an excellent tracer of spiral arms. Finally, the combination of optical sky surveys and DENIS data will improve our knowledge of the distribution of interstellar extinction and the determination of distances.

3. BASELINES OF THE DENIS PROJECT

3.1 The Focal Instrument

The survey will be achieved with a specially designed and dedicated three channel camera equipped with two infrared arrays (JK' bands) and one CCD/(I band). This camera will be attached at the Cassegrain focus of the 1-meter telescope of the European Southern Observatory (ESO) at La Silla, Chile. The telescope has been allocated to this program in the framework of an ESO keyprogram.

The infrared camera consists of a set of three dewars each equipped with one array. The J and K' dewars each house one NICMOS three array (256 x 256) made by Rockwell International and the I band one Tektronix CCD array (1024 x 1024). Both types of detectors are cooled to liquid nitrogen temperature. The f/15 beam of the telescope is split off into three separate beams thanks to two dichroic mirrors. The field of view of the camera is 12'. A microscanning device acting on the J and K' beams allows an improved sampling of the infrared images. The full coverage of one hemisphere will result in about 1 million elementary images per color, taking into account some overlapping. The sky will be scanned in a step and stare mode at constant right ascension, near the meridian, along declination arcs of 30 degrees amplitude, by steps of 10'. It is expected that, taking into account average meteorological conditions at La Silla, the coverage of the negative declination range should be completed in less than four years of observation. Data are handled and processed in real time with three independent controllers based on 68040 Motorola processors. The data flow will be approximately 130 kbytes/sec/channel that will yield an average of eight Gbytes per night and a final amount of data of some four Tbytes. Raw data will be stored on digital audio tapes (DAT) after compression. Sources brighter than approximately K = 8 will saturate the NICMOS arrays. An additional identical filter, but grey-attenuated by five magnitudes, has been added on each channel. The main data process tasks that will be achieved in real time are flat fielding, sky subtraction, astrometric and photometric calibrations and identifications with entries of the Guide Star Catalog. A preliminary small source catalog will be produced in real time in order to allow the immediate follow up of newly discovered interesting sources, but the final data processing will be performed in two dedicated computing centers located at Leiden Observatory and Paris Institut d'Astrophysique.

3.2 Data Processing and Analysis

The two analysis centers will achieve improved standard reductions, each one being in charge of a complementary task in the data reduction stream. The center in Paris will receive the raw image data on DAT, re-process them in a homogeneous way and make them scientifically usable. Elementary images in three colors will be archived and made accessible in a databank. A copy will be forwarded to Leiden where the extraction of point-like objects and the astrometry will be achieved to create a small source database that will eventually become the small source catalog. The Paris center will, in turn, identify and archive the extended sources and will create an extragalactic databank with the participation of Lyon Observatory. Other more specific databases will be implemented under the responsibility of the partners or users, but the normal access to the DENIS data, before final release will be through the two dedicated centers. Data will be delivered in stages to open access to the community before the completion of the survey. However, the co-investigators of the project will have a privileged access to the new data during one year.

3.3. Organization and Planning

DENIS is a collaboration between several Institutes pertaining to seven European countries (Austria, France, Germany, Hungary, Italy, Netherlands and Spain) and a significant contribution of Brazil. The project involves about 60 scientists and engineers. The Space Department (DESPA) of Paris Observatory is leading the project and is responsible for the construction of the three channel camera with important technical and funding contributions of the partners. An operations team of two senior and two postdoctoral astronomers and one dedicated engineer, all resident in Chile will carry on the survey and take care of the continuity of the observations. Commissioning of the camera at the telescope is planned by the end of '93 and the survey should start early in '94. The total cost of the project, excluding salaries, but including three years of operations is approximately 12 MFF (2.4 M$). The project is funded with the contributions of several National Institutions of the partner countries, and of the SCIENCE plan of the Commission of the European Community.

REFERENCES

Epchtein, N., Le Bertre, T. and Lépine, J. R. D. 1990 A&A 227, 82
Epchtein, N. and Mamon, G. A. 1992 Proc. of the 2nd DAEC meeting on the Distribution
    of Matter in the Universe, G. A. Mamon and D. Gerbal, eds., Paris Observatory, p. 382
Habing, H. J. 1988 A&A 200, 40
Guglielmo, F., Epchtein, N., Le Bertre, T., Fouqué, P., Hron, J., Kerschbaum F. and Lépine,
    J. R. D. A&AS 99, 31
Kleinmann, S. 1992 in Robotic telescopes in the 1990s, ASP Conference Series, Vol. 34, A.
    V. Filipenko, ed., p. 203
Neugebauer, G. and Leighton, R. B. 1969 Two Micron Sky Survey, NASA, SP 3047

DISCUSSION

EICHHORN: Have you thought about keeping the data at Paris and Leiden, but make them available through a distributed data access system like ADS or ESIS?

EPCHTEIN: During the on-going period of the survey, the data will be accessible in stages, but only through the data centers in Leiden and Paris, either through ftp or email or, for large amount of data (images) directly at the centers (DAT copies). After survey completion, about five years from now, the end-products should be distributed by the CDS in Strasbourg which will determine the best remote access way available at that time.

PHILIP: This looks like a major effort. How many people are working on it now, and how many when you reach the observing stage?

EPCHTEIN: The DENIS consortium includes about 60 co-investigators and associates, but the core of engineers and scientists deeply involved in the camera implementation and data analysis consists of about 20 persons. We expect to increase the data analysis staff in the coming year to face the flow of data.

CASERTANO: What is the advantage of using the K' filter instead of K?

EPCHTEIN: Simply to reduce the thermal background emission of the sky, instrument and telescope on the array, by cutting the long wavelength part of the K filter. This will allow a gain of 0.5 magnitude in sensitivity with respect to the usual K filter.

THE TWO MICRON GALACTIC SURVEY

P. L. Hammersley, F. Garzón, T. Mahoney, X. Calbet and M. J. Selby

Instituto de Astrofisica de Canarias

ABSTRACT: The Two Micron Galactic Survey (TMGS) is the most sensitive near-infrared large scale survey of the galactic plane yet attempted. The TMGS has so far covered some 255^2 square degrees of sky to a limiting magnitude of $m_k \approx +10$ and detected some 470,000 sources. The majority of these are in the Galactic plane and have no known visible counterparts. The TMGS is also detecting a number of dark nebulae, particularly towards the galactic center.

1. INTRODUCTION

Visible surveys of the galactic plane and bulge are limited, except for a few clear windows, to a maximum heliocentric distance of about three kpc because of high interstellar extinction (Schmidt-Kaler 1976). This limitation may be partially overcome by observing in the near infrared, which suffer considerably less from extinction and in which the flux from stellar sources is still high. To date infrared surveys have had to compromise between sensitivity, spatial resolution and area covered. Most of the near infrared surveys have either been restricted to low-resolution surface brightness mapping (e.g. Okuda 1981) and have consequently been unable to detect individual sources, or they have covered large areas of sky but at a low sensitivity (e.g., the Two Micron Sky Survey, Neugebauer and Leighton 1969), or have only sampled very small areas of sky. The largest high-resolution map is that of Glass et al. (1985), which covered $1° \times 2°$ degrees of the galactic center. The TMGS also covers only a relatively small area of sky, but the area covered is about 200 times more than in previous high-resolution surveys. Full sky coverage with high sensitivity will have to wait for the Deep Near-Infrared Survey (DENIS) and the Two-Micron All Sky Survey (2MASS).

It is expected that M-giants will make up the majority of the sources detected by the TMGS (Harris and Rowan Robinson 1977). Table 1 lists the distance to which the TMGS will be able to see typical sources. It should be able to detect late M giants out to the galactic center (8 kpc) and supergiants throughout the Galaxy, and hence will be able to probe the distribution of sources deep within the Galaxy.

2. THE SURVEY

The TMGS has been previously outlined in Mahoney et al. (1990) and described in detail more recently in Garzón et al. (1993). It is a collaborative project between the Instituto de Astrofisica de Canarias (Spain) and Imperial College (ICSTM, London), which was started in 1988. The TMGS uses the ICSTM 7-channel linear infrared camera on the 1.5-m Carlos Sanchez Telescope (TCS) of the Observatorio del Teide, Tenerife. Being at a height of 2380 m, the TCS is usually well above the main atmospheric water-vapor content and is hence ideally placed for infrared observations. The camera has a pixel size of 15" square and the array is

Workshop on Databases for Galactic Structure
A. G. Davis Philip, Bernard Hauck and Arthur
R. Upgren, eds. p. 103 - 107 © 1993 L. Davis Press

P. L. Hammersley
Instituto de Astrofisica de Canarias
La Laguna, Tenerife
Spain

aligned in declination giving a total field of view of 15" by 115" on the TCS. Under normal conditions the camera has a 1 σ 1 second limiting magnitude of $+12.5$ at K.

TABLE 1

Maximum distances to which the TMGS is expected to detect various source types, using absolute K-magnitudes given in Table 2 of Wainscoat et al. (1992).

Star type	M_K	r for $m_K = 10$ (kpc)
B0 V	-3.00	3.1
K1 V	+4.08	0.15
M5 V	+7.43	0.032
K5 III	-3.10	3.23
M0 III	-4.14	4.60
M3 III	-5.23	6.62
M5 III	-6.90	10.5
M7 III	-9.04	16.0
M3 I	-11.00	24.0
AGB M	-4.38 to -6.80	5.0 to 15
AGB C	-7.41 to -8.83	10.5 to 14
AGB CI	-5.81 to -8.04	7.8 to 21

AGB M, C and CI refer to oxygen-rich, optically visible carbon-rich and optically invisible carbon-rich asymptotic giant branch stars respectively. An extinction of 0.17 magnitudes per kpc is used.

The camera is operated without sky chopping and with the telescope drive stopped, so that the sky is drift scanned at the sidereal rate. A sample rate of 9.361 Hz is used giving significant over-sampling in RA. The scans are typically two hours in length to cover a latitude range $-15°$ $< b < 15°$. By offsetting scans by 105" in declination we are able to build up maps, typically 1.5° in declination by 30° in RA. The main regions covered are at $\delta = -30°$ (running just below the galactic center and including the bulge and disk), $\delta = -23°$ (through the bulge and disk), $\delta = -1°$ and regions. Accurate positioning of each scan is achieved by cross-correlating the positions with the SAO and Guide Star catalogues. Typically the final error in the TMGS source positions is 2" in RA and 5" in declination.

The 3 σ limiting magnitude for the survey is $m_K \approx +10.5$ but as there are gaps between the detectors some of the fainter sources are missed and we estimate the completeness limit is about $m_K \approx +9.8$. By repeating scans and by comparing the TMGS magnitudes with photometry of selected sources we estimate the photometric accuracy is ≈ 0.2 mag, although this is degraded where there are high concentrations of sources.

The relatively large pixel size used means that confusion becomes a serious problem where there are high sources densities. The effects of confusion in the cumulative star counts start at a sources density of about 1000 sources per square degree, but only become severe at about 10000 sources per square degree, where we estimate that up to 50% of the sources at fainter magnitudes are being lost.

3. THE RESULTS

The total star counts per square degree for the areas $\delta = -30°$, $-23°$ and $-1°$ are shown in Fig. 1. As there are a large number of stars in each bin the structure visible is due to the real distribution of sources and not statistical effects. However, since no allowance has been made for confusion, the star counts are reduced where there are very high source densities.

The most noticeable features are the large star densities in the region closest to the galactic center and the pronounced dip in the star counts at RA = 17.7^h, $\delta = -30°$ which is due to the presence of very thick dust close to the galactic plane in the direction of the galactic center. There are also noticeable dips at RA = 18.2^h, $\delta = -23°$ and RA = 18.8^h, $\delta = -1°$, although neither is as prominent as that towards the galactic center.

These curves demonstrate that the TMGS is detecting bulge sources. Both the curves at $\delta = -30°$ (l = 359°) and $\delta = -23°$ (l = 7°) are considerably wider than that at $\delta = -1°$ (l = 30°). Also there are too many stars near the peak of the $\delta = -23°$ and $\delta = -30°$ zones than can be explained by disk stars alone. Furthermore, although both the $\delta = -23°$ and $\delta = -30°$ scans cut through the bulge, at the peak of the $\delta = -30°$ curves there are almost twice as many sources as at the peak of the $\delta = -23°$. Assuming that the local distribution of stars is identical (this is not unreasonable given that the two scans are only eight degrees apart), then the excess must be due to stars within one kpc of the center of the Galaxy. The ratio of the number of stars in each magnitude bin for the peaks of the two areas is presented in Fig. 2. The ratio is about one for stars brighter than $m_K = +5$ and the excess starts at $m_K = +6$. At the distance of the galactic center this would correspond to an M supergiant, one of the most luminous types of infrared source known. The ratio reaches a peak of three at $m_K = +7$ and then begins to decline; however this is probably due to confusion affecting the galactic center region more than the $\delta = -23°$ area. At least 60% of the sources fainter than $m_K = +7$ towards the galactic center are physically close to the galactic center.

In Fig. 3 a map is presented giving the source positions in the region $17.5^h h < RA < 17.9^h$ $-29.2° < \delta < -30.5°$ to a limiting magnitude of +9 mag. This region runs within 0.25° of the galactic center. No attempt has been made to indicate apparent magnitudes. The blank horizontal stripes are due to a the rejection of a few scans for various reasons, whilst these spoil the cosmetic appearance of the picture, they do not seriously effect the results.

Again the most striking feature of the map is the dust lane which makes a Y shape (on closer inspection this becomes an X). In many areas the star counts are reduced to close to zero implying a K extinction of at least three or four magnitudes, and therefore a visual extinction of 30 to 40 magnitudes. Many features already known to be close to the galactic center clearly show up in the TMGS as dark nebulae. Special mention may be made of Sgr C, which is at the fork of the Y and the north western galactic lobe (Sofue 1984) which makes up the branch of

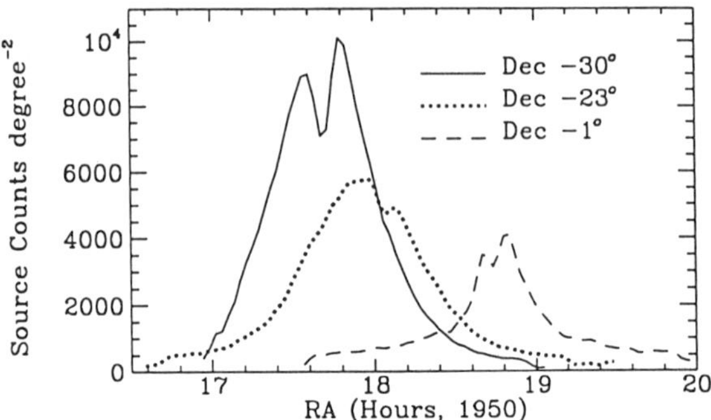

Fig. 1. The total star counts per square degree for the surveyed areas at δ = -30d°, -23° and -1°.

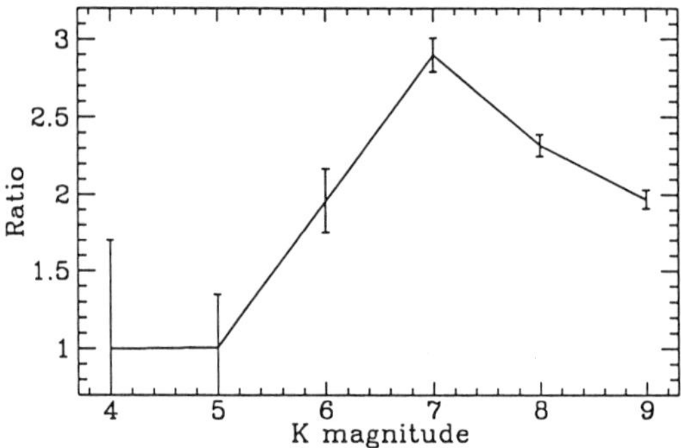

Fig. 2. The Ratio of the star counts at each magnitude for the peak of the curves at δ = -30° and δ = -23°.

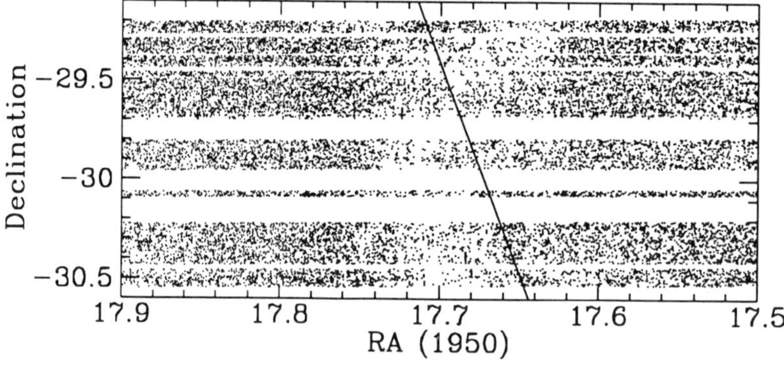

Fig. 3. The map of the region between 17.5^h < RA < 17.9^h -29.2° < δ <- 30.5° for the sources to a limiting magnitude of +9 mag. The galactic plane (b = 0) has also been drawn.

the Y in the direction of decreasing RA. The south western galactic lobe is also visible but it is not as strong. The "trunk" of the Y lies below, at an angle of 30° to the galactic plane. It does not appear clearly on any radio map and is only just visible at IRAS wavelengths, although this is probably due to its being lost in the contours at the edge of the nuclear disk. It is noticeable that the trunk of the Y appears to bend so that it is parallel to the plane at δ = -30.5°. The drop in the star counts in these regions, approaching 80% in small areas, indicates that the proportion of stars physically close to the galactic center is even greater than suggested earlier.

4. CONCLUSIONS

The TMGS is an important database for work on the distribution of sources deep within the Galaxy. It is detecting a large number of sources physically close to the galactic center and a number of dark nebulae, most of which are close to the plane. As well as being a source of information in its own right it will be valuable when planning future observations in the regions covered, particularly with ISO.

REFERENCES

Garzón, F., Hammersley, P., Mahoney, T., Calbet, X. and Selby M.J. 1993 MNRAS, in press
Glass, I. S., Catchpole, R. M. and Whitelock, P. A. 1987 MNRAS 227, 373
Harris, S. and Rowan-Robinson, M. 1977 A&A 60, 405
Mahoney, T., Selby, M.J., Garzón, F., Hammersley, P. and Calbet, X. 1990 in Proc. ESO/CTIO workshop, Bulges of Galaxies, B. J. Jarvis and D. M. Terndrup, eds., ESO, Munchen
Neugebauer, G. and Leighton, R. B. 1969 Two Micron Sky Survey, NASA SP-3047
Okuda, H. 1981 in Proc. IAU Symposium. 96, Infrared Astronomy, D. P. Cruikshank, ed., Reidel, Dordrecht
Schmidt-Kaler, T. 1976 Vistas Astron 19, 69
Sofue, Y. and Handa, T. 1984 Nature 310, 568
Wainscoat, R. J., Cohen, M., Volk, K., Walker, H. J. and Schwartz, D. E. 1992 ApJS 83, 111

FAINT STAR COUNT DATABASES AND GALAXY MODELING

Michel Crézé

Observatoire de Strasbourg

ABSTRACT: Currently accepted ideas concerning the structure and evolution of stellar populations in our Galaxy are mainly based on the detailed observations of stars in the close solar neighborhood. Faint star counts in selected fields have long been advocated only to get an overview of the large scale features. Only recently did modern digitization techniques open the way to getting medium accuracy photometry and accurate proper motions for complete faint star probes in large selected fields. Such data, which can be obtained for many stars, do not allow deriving intrinsic parameters (mass, age, luminosity, gravity) of individual stars. However, some information relevant to the distribution of these quantities is reflected in the n-dimensional distribution of observables. Recent efforts to collect such probes of the Galaxy as well as the related modeling approaches are reviewed. The basic picture of the Milky Way Galaxy as it emerged by 1965 has not been dramatically altered. However important new results were obtained by various groups, including the Thick Disk population, the non existence of a local unseen mass disk or the detection of the edge of the disk. There are still clues of galaxy evolution to be expected.

1. INTRODUCTION

The way we are trying to draw a picture of our Galaxy is reminiscent of the way old geographers used to map remote parts of the Earth: They collected pieces of evidence from sailors back from remote trips and they tried to put those pieces altogether in a coherent picture. Their databases were filled with logbooks and personal notes from travelers. Unknown islands and continents eventually emerged from their careful compilations.

Concerning our Galaxy, the quest started with nearby stars. It was based on systematic observations targeting first bright stars and proper motion stars. The results are very well summarized in the various chapters of "Galactic Structure" (Blaauw and Schmidt 1965). The main features remain basically valid: The differential rotation; the mass distribution deduced from dynamical considerations; the Halo-Disk separation based on both kinematical and chemical properties; the existence of an intermediate population (not yet defined as Thick Disk); the asymmetric drift and the raise of velocity dispersions with age. Some important improvements would follow, concerning the luminosity function (Luyten 1968, Wielen 1974) or the local kinematics or the chemistry, but the basic picture still holds. It was essentially based on observations from the "local swimming hole".

It was not until the seventies that new techniques offered real capabilities to produce complete in situ probes of the remote parts of the Galaxy. New digitization techniques made quantitative photometric and astrometric measurements feasible for complete samples of faint

Workshop on Databases for Galactic Structure
A. G. Davis Philip, Bernard Hauck and Arthur
R. Upgren, eds. p. 109 - 116 © 1993 L. Davis Press

Michel Crézé
Strasbourg Observatory
11 rue de l'Université
67000 Strasbourg, France

stars. The next chapter reviews a number of such attempts conceived as "sample surveys" as opposed to all sky surveys. The review is by no means exhaustive, nor is it intended to be. The emphasize has been put on what is thought to have contributed most significantly to modify the picture, with ample provision for the author's personal biases. There has been in the past a number of contradictory conclusions made out of such or such specific sample. However, provided each contribution is critically reviewed in terms of calibration and statistical significance, provided data are not over interpreted, the general picture that comes out is quite consistent.

2. SAMPLE SURVEYS

The main data providers were Schmidt Telescopes. Bok and Basinski (1964) and the Basle group (Becker 1965 and Becker 1980) gave the early momentum featuring real quantitative photometry. Becker designed a specific photometric system (the RGU system) which passbands were selected to optimize Halo-Disk separation. The Basle program includes fields in many galactic directions and the Basle group was able to derive estimates of the scale height and scale length of the disk as well as the halo density. The picture which comes out from this program is unconventional, the Halo density is very large and the scale length of the disk rather short. However, there are two weak aspects. For one thing there was no photoelectrical implementation of the RGU photometry until the late eighties, and the calibration had to be performed through UBV standards. For another thing, eye-measurements made at the iris photometer could not sample very large areas of the program Schmidt plates. The interpretation based on individual diagnostics of luminosity and abundance for each star, may in some cases result in dramatic systematic errors.

Weistrop (1972), in a survey covering 13.8 square degrees near the NGP, obtained UBV photometry for 14,000 stars. The survey tried to constrain the local space density of faint red main-sequence stars. A population of low velocity red dwarfs which might have escaped the Luyten survey was suggested from the early interpretation. Calibration problems and difficulties in the assignation of absolute luminosities to individual stars may have created uncontrolled problems. Final results did not feature spectacular new results (Weistrop 1979).

Chiu (1980), used prime focus plates from the Lick 3-m and KPNO 4-m telescopes. He was able to reach much fainter magnitudes (BVR, V > 22) in three fields (SA 57, SA 51, SA 68), the areas are rather small (0.1 square degrees) giving no more than a hundred stars in each sample. Not only photometry, but also proper motions were measured. Using long focus plates and pure differential astrometric reduction the author was able to achieve an accuracy quite comparable to the current proper motions in bright star catalogues. The interpretation in terms of Halo and Disk luminosity function was uneasy since again proper motions and colors do not allow unambiguous population and luminosity assignments. Also the samples although reaching magnitude 22 were far to small to add significant constraints beyond the Luyten results. The importance of Chiu's results lies more in the technical break-through in faint star astrometry and photometry.

Jarvis and Tyson (1981), as well as Koo and Kron (1982), investigated small areas on prime focus KPNO 4-m plates. The photometric system is J F. The observing programs were mainly intended (and optimized) to contribute to the statistics of faint remote galaxies. However it showed pretty well how, at high galactic latitudes, the one peak color distribution of stars dominated by the disk main sequence at relatively bright magnitudes, gradually gives place to a two peak distribution at fainter magnitudes. The blue peak implies a substantial non disk population.

Reid and Gilmore (1982), published under the title "New Light on Faint Stars" an analysis of a 18.4 square degrees in VRI at the South Galactic Pole. The main purpose was to provide a sample of the faint red end of the luminosity function on pure photometric grounds. The (V-I) index had been chosen to provide an efficient absolute magnitude criterion for such stars. 12,500 stars have been found down to I = 18.

Although the authors claim to have set an upper limit on the local density of red dwarfs, they actually counted no more than one star fainter than M_v = 14, but due to the uncertainty in the absolute magnitude calibration for such faint objects, even this upper limit is quite uncertain. Furthermore, no new faint red dwarf has been found in addition to those detected by Luyten. To some extent the author's goal to check for the existence or non-existence of an hypothetical low velocity population of faint red dwarfs has been reached and the answer is negative. However the calibration uncertainties at faint magnitudes makes it difficult to assess how far the constraint has been set.

Another important result from the Reid's survey was published by Gilmore and Reid (1983). They gave evidence for the existence of a break in the density distribution of main-sequence red dwarfs beyond 500 pc. Below this limit the distribution is suitably represented by an exponential decrease with scale height about 350 pcs. Beyond the limit, counts are dominated by another population with scale height about 1100 pcs. The distribution of this population is quite distinct from the one of globular clusters. Yet not a fully new entity, this population called "intermediate" in the 1965 picture, has now a well defined identity, including a scale height and an estimate of the local density. Locally the density is ten times larger than the classical halo subdwarf population. There is still room for some criticisms concerning the absolute luminosity assignations made by Gilmore and Reid on the basis of the (V-I) index. This argument was used by Bahcall who claimed that star count data can be suitably modeled (Bahcall and Soneira 1984) using only two populations. However most later investigations based on different samples, and using the population synthesis approach which permits a much safer control of calibration problems, do support Gilmore and Reid's conclusions.

There is still one question left open: Between the free-fall collapse of the proto-galactic cloud and the slow heating of the stellar disk, a new mechanism should be advocated to explain the existence of the Thick Disk.

Stobie and Ishida (1987), published UBV photometry down to V = 19 within 21.46 square degrees around SA 57. The material used was made of Schmidt plates obtained at the Japanese Kiso Telescope. The sample adds up to 18,000 stars. In Yoshi et al. (1987), the authors show how the two color distribution of their sample at different magnitudes cannot be properly modeled unless they use three components including a Thick Disk with intermediate metallicity.

By 1982, the Besançon group developed a new strategy to attack the problem of stellar population synthesis and star count interpretation. This approach is described in three papers: Robin and Crézé (1986), Bienaymé et al. (1987) and Crézé et al. (1989). Most galaxy models are "phenomenological": they are based on empirical luminosity functions. Arbitrary density laws are assumed and scale heights are fitted to star counts.

The Besançon model has been built on the basis of an evolutionary scenario: stars are created all along the galactic disk lifetime. The star formation is governed by realistic (yet arbitrary) star formation rate and initial mass function. Their present luminosity and color result from their initial mass and chemical composition together with evolutionary tracks in the HR diagram. Local velocity distributions are age dependent (based on the observed age/velocity distribution relation and the asymmetric drift). Space distributions are dynamically consistent with velocity distributions. As a consequence the model lends itself to a number of new applications of starcount data.

The dynamical consistency makes it possible to relate star counts to the mass distribution via the potential. Then various hypotheses concerning the unseen mass can be tested. Bienaymé et al. (1986) conclude from a compilation of heterogeneous starcounts in many fields at high and intermediate latitudes that the local mass density which generates the best match with those star counts does not imply any specific local unseen mass. This is in contradiction with claims by Bahcall (1984), who derived local mass estimates twice as large as the observed mass. The force law perpendicular to the galactic plane (the "Kz") obtained in the later investigation was constrained by tracers selected from spectral type surveys (a few stars in the distance range considered). In the Besançon model, due to the fully synthetic approach, general star counts at different latitude can be used, providing thousands of stars in the relevant distance range. The resulting Kz law has been fully confirmed two years later by Kuijken and Gilmore (1989abc) on fully different grounds, the discussion is given in Crézé (1991).

Another specification of the Besançon model is the processing of interstellar extinction which makes it possible to interpret star counts at low galactic latitudes. Mohan et al. (1988) published a Schmidt plate survey covering 3.5 square degrees in a low extinction window near the galactic anti-center (Kapteyn's special selected area 23). The photometry is UBV down to V = 17 the sample includes more than 10 000 stars, photoelectric standards and CCD frames have been used all over the field. A mapping of the extinction has been made and the scale length of the galactic disk is estimated. It is found to be about 2.5 kpc, in close agreement with former results obtained from fitting the Besançon model at intermediate galactic latitude.

The sample survey in Special Area 23 was later completed by much deeper CCD probes obtained at the CFHT 3.6-m telescope in Hawaii (Robin et al. 1992a). The additional survey covers no more than 0.008 square degrees, but it detects stars as faint as V = 26. B, V photometry, better than 8 per cent, is obtained for 400 stars down to V = 25, and UBV down to V = 22. The distribution of the extinction along the line of sight derived from Schmidt plates is confirmed, but over the whole magnitude and color range, all the contributions which would be expected from old main-sequence disk stars beyond 6 kpc vanish although such stars dominate counts at distances less than 5 kpc. This is the signature of a sharp cutoff in the star density: the edge of the galactic disk between 5.5 and 6 kpc. As a consequence, the galactic radius does not exceed 14 kpc. The cutoff of the stellar galactic disk seems to coincide with the region where the H I disk flares away from the plane (Bosma 1978).

More recent star count data include not only photometry but also proper motions. Majewski (1992), follows the line opened by Chiu years before. A new series of plates obtained at the KPNO 4-m telescope extends the time baseline. Proper motions better than 0.2 m.a.s. per year and UBVI photometry have been obtained for 250 stars down to B = 22.5. Majewski shows some kinematical evidence for a discontinuity in the kinematical properties of stars beyond 1700 pc above the galactic plane, which he interprets as the kinematical signature of the end of the Thick Disk. His conclusions complete the description of the Thick Disk although there is still room for criticizing the size of his sample and the calibration of individual distances (dwarf/giant separation).

Bienaymé et al. 1992 use recent Schmidt plates in combination with glass copies of the first Palomar Sky Survey, thus obtaining a time baseline larger than 35 years. The astrometric reduction is fully differential and uses orthogonal polynomials to get the plate to plate transform. Proper motions are better than 0.3 mas per year and UBV photometry is obtained for 1180 stars in two square degrees at intermediate galactic latitude. The zero point of proper motions is set by comparison with the zero point predicted by the Besançon model.

Soubiran 1992, adopts Bienaymé's method, to investigate a field close to the North Galactic Pole. The set of plates was completed by a series of plates obtained at the Tautenburg Schmidt Telescope. Observing epochs are well spread all along the time baseline, thus providing an external test of the validity of the Bienaymé's astrometric method. The final accuracy is better than 0.2 mas for 4000 stars over seven square degrees. B, V photometry has been obtained for 18,000 stars in 20 square degrees. The zero point of the proper motion is fixed by reference to galaxies in the field.

The full interpretation of the last survey is not yet finished, but there is already one result: the fit of proper motion distributions at different magnitudes with those predicted by the Besançon model is excellent, although there has been no fitting of the model on those data, (apart from the zero point in the former sample). So this is a very good validation of local differential astrometry for proper motions. There are also preliminary indications concerning the rotation of the Thick Disk.

Combined photometric and proper motion samples provide five-dimensional data cubes for in situ probes of the galaxy outside the solar neighborhood. Such data, which can be obtained for many stars, do not allow deriving intrinsic parameters (mass, age, luminosity, gravity) of individual stars. However, some information relevant to the distribution of these quantities is reflected in the five-dimensional distribution of observables. Early attempts to extract this information by eye comparison in two-dimensional marginal diagrams are far from exhausting this information. There are still methods (both parametric and non-parametric) to be developed to cope with such multivariate problems. The preliminary investigations we have made show that there are promising possibilities. It is potentially a powerful source of evidence for the evolutionary processes which made our galaxy look like it does.

3. CONCLUSIONS

Sample surveys suitably combined with galaxy modeling have already produced a number important results:

- First of all they have definitely validated on large scales the main inferences about stellar distributions in our galaxy which had been made before 1965 on the basis of very local data. However non exciting it may look this result was not at all trivial.
- Among the new features that emerged, is the identification of the intermediate population as a "Thick Disk" with well separated properties in terms of space density, kinematics and chemical composition.
- Another new feature is the edge of the galactic disk at 5.5 kpc away from the galactic plane.
- Unexpectedly, faint star count data can also be credited an accurate determination of the Kz force law which rules out the existence of large amounts of unseen mass in the galactic disk.

Although one of the main motivations of most photometric surveys at high latitude, our knowledge of the faint red part of the luminosity function got very little benefit from those investigations. The reference is still the Luyten or Wielen Luminosity Function. And our best hope to get substantial improvement in the faint part of the distribution still relies on all-sky proper motion surveys. The only contribution of sample surveys in this field has been to rule out very dense ad hoc populations of low velocity very faint red dwarfs. Multivariate sample surveys including proper motions magnitudes and colors potentially bring a new insight in evolutionary processes in the early epoch of the galactic disk, their interpretation is still just beginning.

REFERENCES

Bahcall, J. 1984 ApJ 287, 926
Bahcall, J. and Soneira, R. M. 1984 ApJS 55, 67
Becker, W. 1965 Zs.Ap 62, 54
Becker, W. 1980 A&A 87, 80
Bienaymé, O., Mohan, V., Crézé, M., Considère, S. and Robin, A. C. 1992 A&A 253, 389
Bienaymé, O., Robin, A. C. and Crézé, M. 1987 A&A 180, 94
Bok, B. J. and Basinski 1964 Mem. Mt. Stromlo Obs. 16
Chiu, Y. -T. G. 1980 ApJS 44, 31
Crézé, M. 1991 in IAU Symposium No. 144, The interstellar Disk-Halo Connection in
    Galaxies, Reidel, Dordrecht, p. 313
Crézé, M., Bienaymé, O. and Robin A. 1989 A&A 211, 1
Gilmore, G. 1981 MNRAS 195, 183
Gilmore, G. and Reid, I. N. 1983 MNRAS 202, 1025
Jarvis, J. F. and Tyson, J. A. 1981 AJ 86, 476
Koo, D. C. and Kron, R. 1982 A&A 105, 107
Kuijken, K. and Gilmore, G. 1989a MNRAS 239, 571
Kuijken, K. and Gilmore, G. 1989b MNRAS 239, 605
Kuijken, K. and Gilmore, G. 1989c MNRAS 239, 651
Luyten, W. 1968 MNRASoc 139, 221
Majewski, S. R. 1992 ApJS 78, 87
Mohan, V., Bijaoui, A., Crézé, M. and Robin, A. C. 1988 A&A 73, 85
Reid, N. and Gilmore, G. 1982 MNRAS 201, 73
Robin, A. C. and Crézé M. 1986 A&A 157, 71
Robin, A. C., Crézé M. and Mohan V. 1992a A&A 265, 32
Robin, A. C., Crézé M. and Mohan V. 1992b ApJL 400, 25

Sandage, A. R. 1987 in The Galaxy, G. Gilmore and R. F. Carswell, eds, Reidel, Dordrecht,
    p. 321
Soubiran, C. 1992 A&A 259, 394
Stobie, R. S. and Ishida K. 1988 AJ 93, 624
Weistrop, D. 1972 AJ 77, 366
Weistrop, D. 1979 PASP 91, 193
Wielen, R. 1974 Highlights Astron., 3, 395
Yoshi, Y., Ishida, K. and Stobie, R. S. 1987 AJ 92, 323

DISCUSSION

ROMAN: Is the 5.5 kpc cut-off sharp or is this a mean distance of a more gradual decrease?

CREZE: It is a sharp cut-off: the density falls down to nearly nothing within a few hundred kpc.

ROMAN: Can the apparent cut-off result from an error in the magnitude scale (e.g., a non-linearity)?

CREZE: Actually the magnitude scale is as good as the linearity of the CCD, of course we have no external calibration at such faint magnitudes. However I would not expect a departure from linearity of the size needed to produce such an effect. On the other hand, there is convergent evidence from the color distribution.

ROMAN: In Majewski's data, can he distinguish clearly between giants and dwarfs? The 30 kpc distance for some stars is strange. If he calls a dwarf a giant, he would get both distances and velocities which are too large.

CREZE: I am not sure what Majewski can actually distinguish. Assigning individual distances to individual stars on the basis of broad band photometry, he is likely to misclassify some giants. However I have little doubt that the evidence he gives for the Thick Disk-Halo discontinuity is real.

RATNATUNGA: You talk about the data being five dimensional. It is clear that only two can be handled graphically, although you can do three or more numerically. What have you done for numerical analysis of stellar data?

CREZE: We are currently developing algorithms based either on multivariate analysis or on pattern recognition techniques; both are of course numerical, but may help tracing model-data discrepancies, better than blind fitting techniques. Such methods are necessary in the exploratory period of the investigation.

RICHMOND: You mentioned that the "sample surveys" do not go deep enough to reveal the faint end of the luminosity function. How deep would surveys have to go?

CREZE: CCD surveys over a small field of view - say, tens of square degrees - down to 25th or 26th magnitude might do it. But we really need proper motions, too, to answer the question.

LOCKMAN: Your results, and those of Kuijken and Gilmore, imply that there is no longer a "missing mass" problem in the solar neighborhood. Does Bahcall, who reaffirmed the existence of the problem in the late 1980s, agree with your conclusion?

CREZE: To the best of my knowledge, John Bahcall does not accept this conclusion yet. But I have not seen any explanation as to how our work and the one from Kuijken and Gilmore, based on fully independent approaches and fully independent data might have produced consistently wrong force laws.

LU: Regarding the disagreement between Bahcall's missing mass and Gilmore's no-missing mass models, Gould, Flynn and Bahcall have recently reconfirmed the missing mass scenario using Flynn's high luminosity radial velocity data and argued the untested velocity ellipsoidal model of Gilmore et al. A model of no missing mass results.

CREZE: The Flynn sample according to Gould's own calculations is too scarce at low z to provide a real constraint on the total mass density in the Galactic plane. By the way, there are recent results by Burkart Fuchs at Heidelberg suggesting that the Bahcall's analysis might be biased due to the fact that the sample tracers they used are not isothermal.

Discussion II. (Moderator A. Omont) Which Kinds of Surveys Will Be Most Useful for Those Who Build Galactic Models?

MODERATOR: To introduce the discussion let me slightly enlarge the topic and briefly recall, as an outsider, a non exhaustive list of problems to be addressed in the field of galactic structure and evolution. Most of them were introduced by previous speakers.

1. The major problem is certainly the nature of dark matter. The details of its mass distribution have to be traced by a complete determination of the gravitational potential, both "locally" and in distant parts of the disk (warp and extent) and of the halo (shape and extent. Details of the mass distribution should be confirmed in the central regions (bulge, bar, etc.), as well as the perturbations in the spiral arms and in the vicinity of giant molecular clouds. Accurate distance calibration is important in this field as in many others.

2. A closely related topic is stellar distribution and dynamics. The limits of the stellar disk and halo are important questions. Extending studies to distant regions to make systematic studies of stellar kinetics is an example of major progress. Examples of questions to be addressed are motions in the bulge and the essential one of understanding the heating of the disk stellar distributions.

3. The main bulk of questions for galactic models remain that of stellar populations, their luminosity function, IMF and evolution. Comprehensive studies of various classes of stars must still be pushed, especially for low-metallicity populations. This includes: detailed studies of open clusters, as well as of globular clusters in relation to distant field stars, radial variations of abundances and of stellar populations, census of distant white dwarfs and of pulsars, etc. We heard that the advent of sensitive infrared surveys will strongly improve our knowledge of many classes of stars: distant disk populations, low mass nearby stars (red dwarfs and hopefully brown dwarfs), infrared stars (young, AGB, etc.).

4. The questions related to the gas distribution and dynamics have not been addressed in the first part of this meeting. Tracing the structure and the physics of the various components (atomic, molecular, ionised, "coronal") of the interstellar gas is important not only from the "geographical" point of view, but also to understand its past and present evolution, including in particular star formation, interaction with young stars and supernovae, exchanges with the intergalactic medium, etc. Closely related topics, particularly relevant for the whole Galaxy, are extensive studies of the interstellar magnetic field and of the interstellar extinction.

In regard to surveys in progress, let me first try to give a non-exhaustive summary of surveys already carried out or to be completed in the coming years:

1. Surveys of the whole sky or of the galactic disk

• Visible: As discussed in this meeting, this is the richest source of information concerning the number of sources and their intensity. Corrections for the extinction can be improved, but remain difficult. Confusion is also a major problem in the disk. Waiting for dedicated digital surveys, existing Schmidt plates contain a huge amount of information. The different operations of digitalization in progress are producing huge digital catalogs in several colors. However,

Workshop on Databases for Galactic Structure
A. G. Davis Philip, Bernard Hauck and Arthur
R. Upgren, eds. p. 117 - 121 © 1993 L. Davis Press

digitalization of the galactic disk (and bulge) areas should be especially addressed, along with the problem of confusion, in connection with infrared surveys.

• Hipparcos will give a fundamental reference frame for parallaxes, and a more limited one for proper motions which are provided in a more sensitive and systematic way from photographic surveys, either general (such as POSS I + II) or dedicated (Lick, Yale/San Juan, etc.).

• The coming near infrared surveys (DENIS amd 2MASS) will be complementary, in visible and IRAS surveys for cold stars, and will extend much deeper into the disk.

• Although not very sensitive, IRAS has provided a complete view of the far infrared sky (very cold luminous stars: AGB, PN, etc.).

• ROSAT is providing the first extensive survey of X-ray sources (important e.g. for low mass stars).

• Various radio surveys cover most of the galactic disk for the different components of the gas at various wavelengths (cm continuum, H I 21 cm, CO lines, etc.) and also for peculiar sources such as supernova remnants, pulsars, OH/IR stars, etc.

2. Deep Surveys of Selected Areas

Deeper studies of limited areas are quite important to address many questions. We heard of several of them. It is clear, as often suggested in the past, that one should try to coordinate such studies in the future to have complete and deep views of the same area at multiwavelengths.

I would like to stress the amount of data accumulating on two regions of special interest: 1) the directions close to the galactic center for obvious reasons ; 2) the line of sight of the Magellanic Clouds, in particular from the operations (MACHOS, EROS) looking for microlensing effects by massive objects such as brown dwarfs of the galactic halo.

3. Which Kinds of Future Surveys?

Many have already been mentioned by various speakers, and/or are obvious extensions of existing surveys, e.g.:

• Digital survey of (parts of) the galactic disk, similar to the Sloan survey.
• Extension of proper motion measurement to later epochs and/or with digital surveys.
• Hipparcos II?
• More systematic surveys of radial velocities.
• Near infrared surveys: deeper in selected areas; repeat at a second epoch; use them for astrometry; etc.
• Future space infrared surveys:
• What to do with ISO, HST/NICMOS, SIRTF?
• Near infrared spectroscopic survey?
• 3 - 8 μm AFGL MSX survey of the galactic disk?
• Others?

4. Some Prescriptions for the Best Utility of Surveys

• Good interaction of users (model makers of galactic structure and other problems) with catalog makers.
• Data conviviality: develop software for use of catalogs and data bases. Standard formats and dedicated cross-identification procedures to allow cross-fertilization of surveys.
• Organization of the selection of specific areas for deeper surveys.
• Model makers have to think how to use such a huge amount of data, including by-products of other purpose surveys (e.g. cosmology).

COHEN: I can advertise for my own talk now! I will show a set of wavelengths selected for the study of galactic structure with ISOCAM. They were chosen because they offer a very wide color separation of galactic sources in the color-color plane that they define.

GRIFFIN: You have already answered the question, "which databases are most useful for Galactic Structure" by enumerating just about every database and survey that one can think of. That highlights the truth that, to some extent, <u>all</u> surveys and databases (except those which are specifically extragalactic) are relevant here. I should also like to remark that the most useful databases (for <u>any</u> sort of study) are those which are as complete and homogeneous as possible and have been adequately vetted and evaluated by (a) knowledgeable entrepreneur(s).

HUMPHREYS: It is obvious from your summary that once these deeper IR surveys are available there will be no comparable optical survey for the galactic plane. Degradation of plates in the plane is possible but good image deconvolution is difficult in such crowded fields. The best approach may be a digital (CCD) survey at a relatively bright limit $\sim V = 15 - 16$ mag.

LASKER: I agree that the low latitude plates are difficult. Some are so crowded as to be confusion limited (and therefore not too useful). The others can be (and to some extent have been) processed, although to a limit no as faint as the high latitude plates, mainly because of a cluttered background. I feel that good work in low latitude fields where the extinction is not too great can be coordinated with the requirement of the infrared programs. Also I call attention to the short exposure plates used on the GSC and to those taken in the red with the UK Schmidt.

MERMILLIOD: In connection with the need for radial velocities, I would like to recall that there is an ESO key program for observing late-type stars from the Hipparcos survey. It contains no less than 25,000 stars in surveys made by Carney and Latham and by Lindgren and Andeberg.

RATNATUNGA: I have a number of comments. First, with respect to the need of a CCD survey on the galactic plane, I understand the bright time of the survey is not subscribed and available to any group with funds to use this time. Second, I wish to give publicity to the IAS Galaxy Model (Ratnatunga, Bahcall and Casertano 1989 ApJ). This model includes stellar kinematics and is available to anyone interested in doing comparison of proper motion and line-of-sight velocity distribution in addition to color and apparent magnitude. Third, with respect to what kind of survey will be most useful for those who construct Galaxy models the most important criteria are (1) automatic survey data, since selection are simpler to model than for sample selected by eye, (2) clear definitions of sample selection and instrumental detection limits, and (3) reliable rms error estimates for individual observables rather than no errors as we unfortunately see in many astronomical catalogs of mean errors for the catalog or a whole which is not useful for detailed statistical analysis. Finally, the Luyten proper motion catalog with $\mu > 0.08"/yr$ is an automated catalog which has not been made available to the astronomical community since it was created over a decade ago. I appeal to Nancy Roman, now at ADC and formerly at NASA, who funded this research and Roberta Humphreys who is at Minnesota where the measurements were made and the data tapes exist, to make all attempts to obtain this catalog. I understand that Luyten is also very eager to see this catalog published.

ROMAN: Schmidt plates are unsuitable for deep surveys in the Milky Way because of confusion. However, it should be possible to get deep survey data in the Milky Way using large scale telescopes. The fields will be small but the star density is large enough that it should be possible to get good statistics.

CASERTANO: In a perfect world, we would like to have surveys that are complete, unbiased, and free from systematic effects. Of course, that is not always possible. I would like to point out that a proper method of analysis can overcome many kinds of incompleteness and systematics. From my point of view, even less-than-perfect surveys can be extremely useful to construct and constrain galaxy models.

LASKER: As we've been discussing data sets in five dimensions it seems natural to ask about the sixth, radial velocity. These are probably the most costly data we'll want to acquire. Can we make good use of the other data to make best use of the resources for radial velocities?

RATNATUNGA: With respect to selecting subsample for example for measurement of line-of-sight velocity, I would suggest selecting smaller areas which could become smaller at fainter magnitudes rather than selecting randomly from larger areas the selection of which is not always clear for analysis.

CREZE: Obviously, one major return from all sky surveys as far as stellar populations is concerned has to be ensured by combining infrared data with proper motion data. Couldn't we think of minimum coordination at the level of early data organization which would make the cross identification much easer?

PLATAIS: I would like to make a comment on large proper motion surveys, namely, Yale/San Juan and Lick, which are nearing completion. Both surveys will provide invaluable and needed data to the galactic structure studies. However, problems related to photographic astrometry, i.e., a limited number of stars to be measured, inevitable systematic errors. I think a true improvement will be possible only after the second generation Hipparcos mission - the Roemer astrometric satellite. Though, I haven't heard a word about it. Why not?

LASKER: As we build relations between data for the same objects from many different sources, the matching becomes a critical operation. A clear step is the assignations of names. We all need to go to a consistent naming convention. To the extent that the astrometry is well done, the IAU naming convention may be the obvious choice.

ROMAN: Is not the real problem with deep surveys in the plane extinction?

LATTANZI: I'm afraid that I don't see a HIPPARCOS II happening before (but estimate) the year 2010. This doesn't mean that we should not help the effort toward making an HIPPARCOS II. Quite the contrary!

OMONT: I am not sure that much has to be added in the way of concluding remarks, but an obvious recommendation is to find the best way to enforce all the suggestions made in this discussion, in particular: Coordinate the various surveys by instauring some kind of network between survey makers and data bases; stress the importance of data homogeneity and providing error estimates; and also of software development and of source identification (and name assignation); organize the selection of areas for deep surveys; publish the Luyten proper motion catalog; push the measurements of radial velocities; take full advantage of Hipparcos results, and try to accelerate the advent of Hipparcos II. However, perhaps the most important conclusion emerging from this discussion is the need for complete visible surveys of the galactic disk and, in particular, for the organization of a digital survey.

A PROGRAM FOR GALAXY POPULATION STUDIES

V. Straižys

Union College

R. P. Boyle

Vatican Observatory Research Group

A. G. Davis Philip

Union College and Van Vleck Observatory

1. MODELS OF THE GALAXY

During the last decade, several models of the Galaxy have been proposed. The most frequently used is the Bahcall and Soneira (1980, 1984) model which takes into account a Pop I disk and a Pop II spheroid. The authors show that their simple model gives a sufficiently accurate picture of the observed distributions of stars in magnitudes and colors in certain directions, all of which have galactic latitudes $> 20°$ (Bahcall 1986). This has been done for magnitudes and color indices in different broad-band photometric systems, and in all cases the observed and predicted surface densities of stars coincided within the expected errors. The model has not been tested near the galactic plane nor in the central bulge.

The Gilmore and Reid (1983) and Gilmore (1984) model contains a thick disk of lower density in addition to a thin disk and a spheroid. The Robin and Crézé (1986a, b) model also has an intermediate population. The Wainscoat et al. (1992) and Cohen (1993) model represents the galactic disk, central bulge, ring, spiral arms and spheroid population. Haud and Einasto (1989) propose a model with a massive corona in which stars probably are not very abundant. More details about the Galaxy models currently in use are given in the reviews by Bahcall (1986) and Gilmore et al. (1989).

The assumed stellar density law for the disk in the Bahcall and Soneira model is following:

$$n(D) = n(D, R_o) \exp[-z/H(M_v)] \exp[-(x-R_o)/h].$$

Here $n(D)$ is the star density at a distance x from the galactic center and at a distance z from the plane of the disk, R_o is the solar distance from the galactic center, $n(D, R_o)$ is the star density in the solar neighborhood (luminosity function), $H(M_v)$ is the vertical disk scale height, and h is the radial disk scale length. Bahcall and Soneira used $R_o = 8.0$ kpc and the local density of stars $n(D, R_o) = 0.13$ stars/pc^3. This value does not include stars fainter than $M_v = 16.5$ mag. The luminosity function, the scale height and the scale length are considered to be

Workshop on Databases for Galactic Structure
A. G. Davis Philip, Bernard Hauck and Arthur
R. Upgren, eds. p. 123 - 130 © 1993 L. Davis Press

V. Straižys
Institute for Theoretical Physics
  and Astronomy
Gostauto 12
Vilnius 2600, Lithuania

the same everywhere in the disk.

The density of the spheroid stars as a function of the galactocentric distance was assumed to follow the exponential law. Since the density of the spheroid in the solar vicinity is only 1/500 of the disk density, we shall neglect these stars when discussing the disk population.

The following values of the vertical scale height of the disk were taken: 90 pc for O-B-A stars with $M_V < 2.3$, 325 pc for G5-K-M main-sequence stars with $M_V > 5.1$, and for F and early G main-sequence stars the scale height was linearly interpolated between 90 and 325 pc; for disk giants and white dwarfs H = 250 pc was assumed. However, a glance at Fig. 2 of the Bahcall and Soneira (1980) paper shows a great uncertainty in the scale height values for different types of stars. There are gaps, interpolations and extrapolations in it. Robin and Crézé (1986a) have used different scale height values, estimated from the spatial velocity dispersion related with age. In addition, Kent et al. (1991) found evidence that the scale height for K and M giants is not constant but increases with radius at galactocentric distances > 5 kpc. It is not known how the scale height behaves at the edge of the disk where the interstellar gas, and probably stars, show a warp (Carney and Seitzer 1993).

For the radial scale length of the disk, h = 3.5 kpc was assumed. This value is estimated from the surface brightness of the Milky Way, from H I density observations in the Galaxy, and from the mean density profile of external Sbc galaxies. However, later determinations of this quantity give very ambiguous results ranging between 1.8 kpc to 6 kpc (see reviews by Kent et al. 1991 and Robin et al. 1992a). The scale lengths of spiral galaxies may be uncertain by a factor of two (Knappen and Kruit 1991). Mohan et al. (1988) and Robin et al. (1992a,b) are in favor of a low value of h = 2.5 kpc which is based on star counts in the galactic plane in the anticenter direction. Kent et al. (1991) conclude that the measured scale lengths are so much different due to some assumptions used in different methods which may not be valid. Also, different components of the Galaxy must have different radial distributions. Consequently, the direct determination of scale lengths for different types of stars is very important.

In their second paper Bahcall and Soneira (1984) decided to use the Wielen (1974) luminosity function for the solar neighborhood. Is the luminosity function the same everywhere in the disk? In his classical papers on the luminosity function McCuskey (1965, 1966) showed variations of the luminosity function with distance from the Sun and perpendicular to the galactic plane (see also Upgren 1963, Philip 1967 and Gilmore and Reid, 1983). Theoretically, a unique luminosity function is not expected, since the disk contains regions differing in age and metallicity. For example, spiral arms contain younger stars than the interarm regions. Also, there is a well-known metallicity gradient, i.e. decrease of metallicity with distance from the galactic center. A different luminosity function is expected in the thick disk (if it is real) and the central bulge, which have a different evolutionary history.

2. A PROJECTED NEW SURVEY

For the determination of luminosity functions or space density functions of stars of different MK types in different galactic longitudes and at different distances from the Sun, for more precise determination of the scale height and the scale length of different types of stars, for the solution of the thick disk problem, and for the verification of the galactic models at low galactic latitudes, a new observational survey near the galactic plane is necessary. To overcome the interference by interstellar dust, the survey must be based on the reddening-free two-dimensional classification of stars. The best photometric method which can be used for this is medium-band CCD photometry of stars in the system which makes it possible to classify stars completely photometrically and in the presence of interstellar reddening. Such is the seven-color Vilnius photometric system. The combination of the Vilnius photometric system with CCD detectors of high sensitivity and large field gives a unique possibility of investigating the galactic disk up to great distances from the Sun. No other photometric system can do this job. The characteristics of the Vilnius system bandpasses are given in Table 1. The bandpasses themselves, normalized to 100 at maximum transmission, are shown in Fig. 1 (with the bandpasses of the four-color sytem for comparison). The system has been described in detail by Straižys (1973, 1977, 1992).

TABLE 1

Mean wavelengths and half-widths of the Vilnius
photometric system

		U	P	X	Y	Z	V	S
λ	nm	345	374	405	466	516	544	656
$\Delta\lambda$	nm	40	26	22	26	21	26	20

The positions of the bandpasses, plotted aside the flux distribution curves of the A0 V and K5 V, are shown in Figs. 2 and 3. The U bandpass measures the intensity to the ultraviolet side of the Balmer jump. The P bandpass measures the absorption of the higher members of the Balmer series which is sensitive to surface gravity in B-A-F stars. The X bandpass measures the intensity to the red side of the Balmer jump for early-type stars as well as the amount of blocking by metal lines for late-type stars. The Y bandpass is placed near the break point of the interstellar extinction law and, in combination with other bandpasses, is important in distinguishing the temperature and interstellar reddening. The Z bandpass is placed on the wide absorption feature formed by the MgI triplet lines and a MgH molecular band. The depth of this feature is very sensitive to surface gravity or luminosity in K and M stars. The V bandpass is a medium-band analogue of V in the UBV system. The S bandpass is placed on the Hα line and plays a role of its emission discriminant among early-type stars. In late-type stars the S bandpass measures the intensity of the pseudocontinuum.

The Vilnius system allows one to classify stars of all temperatures in spectral classes (or temperatures) and absolute magnitudes (or surface gravities) in the presence or absence of interstellar reddening. For the classification of stars no information from stellar spectra is needed. The simplest method of classification is the use of interstellar reddening-free Q,Q diagrams calibrated in spectral types and absolute magnitudes or in temperatures and gravities.

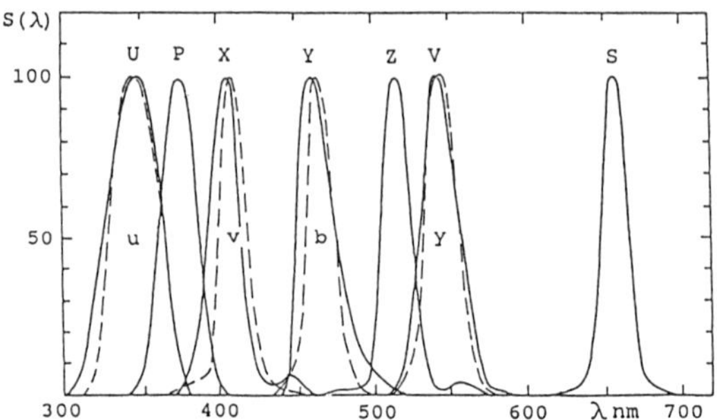

Fig. 1. Response functions of the Vilnius photometric system. This and Figs. 2 and 3 are taken from Straižys (1992). In each of the figures the bandpasses of the Strömgren system are shown for comparison.

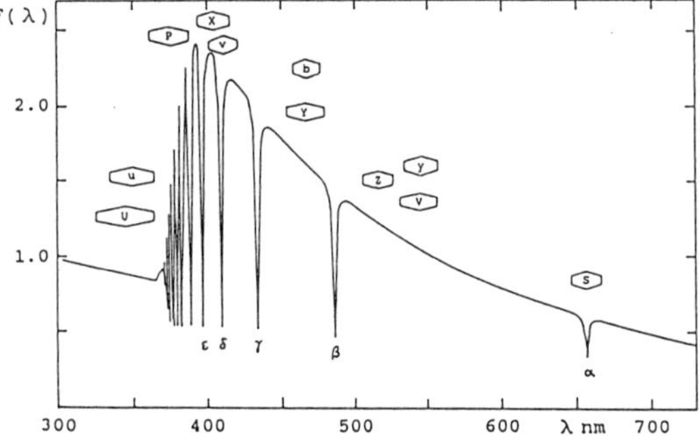

Fig. 2. Energy distribution in the spectrum of α Lyrae with positions and halfwidths of the Vilnius system bandpasses.

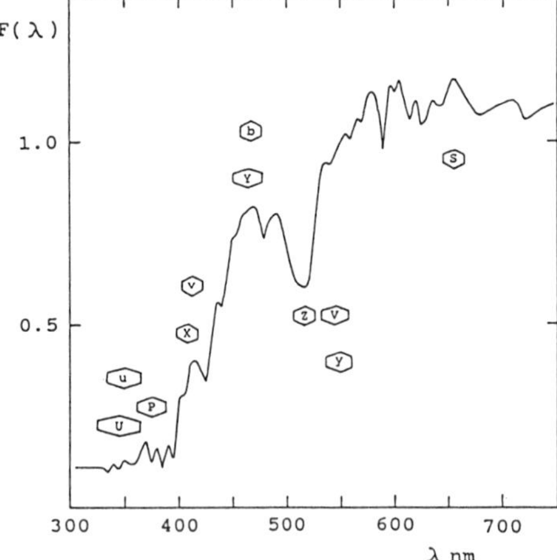

Fig. 3. Energy distribution in the spectrum of a K5 V star with postions and halfwidths of the Vilnius system bandpasses.

The definition of a photometric Q-parameter is as follows:

$$Q_{1234} = (m_1 - m_2) - (E_{12}/E_{34})(m_3 - m_4),$$

where $m_1 - m_2$ and $m_3 - m_4$ are two color indices and E_{12} and E_{34} are the corresponding color excesses. These diagrams and other classification methods are described in the monograph "Multicolor Stellar Photometry" (Straižys 1992).

The system also makes it possible to recognize photometrically the following peculiar types of stars: Be, extreme Am and Ap/Bp, F-G-K subdwarfs, metal-deficient giants and subgiants, carbon and barium stars, Herbig Ae/Be stars and T Tauri-type stars, white dwarfs, and a number of types of unresolved binaries. No other photometric system is capable of solving such a wide variety of classification problems, and in the presence of interstellar reddening. The Vilnius photometric system enables one to classify about 90-95% of stars in any galactic field where stars of different spectral classes, luminosities, metallicities, peculiarities and interstellar reddenings are mixed together. The remaining 5 or 10% of stars which cannot be recognized and classified photometrically, may be very interesting objects for spectroscopic investigation.

Especially great possibilities appear for investigation of star fields when the Vilnius system is used with CCD detectors. The first results of CCD photometry in the Vilnius system have been published by Boyle et al. (1990a, b, 1992) and Smriglio et al. (1991), giving classification of stars down to 15-17 mag. The CCD frames were obtained with an RCA 316 x 508 chip on the 0.9-m telescope at KPNO, with a field size of 5 x 7 arcmin. The exposure times were 20 min for two ultraviolet filters and 2-4 min for the other filters. This means that with exposures of the order of 1 hour for ultraviolet magnitudes it is possible to classify stars down to 18th mag with a 1-meter telescope or to 21st mag with a 4-meter telescope. A potential increase of limiting magnitude can be achieved by using a more ultraviolet sensitive CCD and more transparent ultraviolet filters.

3. THE SELECTED AREAS

For the investigation of the space distribution of stars close to the galactic plane (for this project and for its future extensions) several tens of areas have been selected in the most transparent regions of the northern Milky Way with galactic latitudes $|b| < 10°$ and galactic longitudes l between 6° and 220°. The positions of the areas are based on visual examination of the stellar background within the zone of avoidance in the blue and red copies of the Mount Palomar Observatory Sky Survey. The approximate location of the centers of 25 areas chosen for the present project are given in Table 2. Each of these areas is larger than 0.5°, so there is sufficient space to select smaller fields of about 20 arcmin diameter for CCD photometry.

The results of two-dimensional classification and distance determination of stars down to 18th mag will be used to obtain spatial densities of stars of different types, luminosity functions, vertical scale heights and radial scale lengths in different directions of the galactic plane. This information will be used to verify the existing models of the Galaxy near its plane. A by-product of this program will be an "interstellar extinction vs. distance" relation in the investigated areas, as well as the identification of hundreds of stars of peculiar types.

TABLE 2

List of the transparent areas for the galactic
structure study at low latitudes

Area Nearby object	RA (1950)	DEC	l	b
1. Small Sgr Cloud (OCl M 24)	18 15	-18.0	13.3	-1.1
2. North of GCl IC 1276 (Ser)	18 12	- 5.6	23.6	+5.7
3. GCl NGC 6712 (Sct)	18 50	- 8.8	25.5	-4.5
4. South of N Aql 1918	18 52	- 1.0	32.5	- 1.1
5. At ζ Aql	19 01	+14.2	46.8	+ 3.9
6. West of γ Aql	19 33	+11.3	48.0	- 4.3
7. At OCl NGC 6802 (Vul)	19 27	+19.7	54.8	+ 0.9
8. GCl M 71 (Vul)	19 52	+18.6	56.9	- 3.8
9. GCl M 56 (Lyr)	19 15	+30.1	63.4	+ 7.6
10. In SA 64 (Cyg)	19 58	+30.6	67.8	+ 0.4
11. In the Veil Nebula	20 50	+31.2	73.6	- 7.5
12. At OCl IC 1311 (Cyg)	20 07	+40.8	77.9	+ 4.0
13. Above North America Neb.	21 02	+46.2	87.8	- 0.4
14. Above OCl NGC 7245 (Lac)	22 10	+54.8	101.4	- 1.0
15. South of OCl M 52 (Cas)	23 24	+60.9	112.9	+ 0.0
16. North of Gamma Cas	00 47	+64.0	122.9	+ 1.5
17. OCl NGC 1027 (Cas)	02 38	+61.3	135.9	+ 1.5
18. At ψ Per	03 31	+48.0	149.6	- 6.1
19. OCl NGC 1528 (Per)	04 12	+51.2	152.0	+ 0.3
20. West of 5 Cam	04 42	+55.0	152.0	+ 6.0
21. At ζ Aur	04 58	+41.3	164.7	- 0.3
22. Anticenter	05 43	+30.0	179.2	+ 0.8
23. West of γ Gem	06 22	+16.1	195.7	+ 1.7
24. East of OCl NGC 2244 (Mon)	06 36	+ 4.7	207.4	- 0.7
25. South of 19 Mon	07 02	- 6.2	220.0	+ 0.0

REFERENCES

Bahcall, J. N. 1986 Ann. Rev. Astron. Astrophys. 24, 577
Bahcall, J. N. and Soneira, R. M. 1980 ApJS 44, 73
Bahcall, J. N. and Soneira, R. M. 1984 ApJS 55, 67
Bok, B. J. 1937 The Distribution of Stars in Space, Chicago Univ. Press
Boyle, R. P., Smriglio, F., Nandy, K. and Straižys, V. 1990a A&AS 84, 1
Boyle, R. P., Smriglio, F., Nandy, K. and Straižys, V. 1990b A&AS 86,395
Boyle, R. P., Dasgupta, A. K., Smriglio, F., Straižys, V. and Boyle, R.P., Dasgupta, A.K.,
    Smriglio, F., Straizys, V. and Nandy, K. 1991 A&AS, 95, 51
Carney, B. W. and Seitzer, P. 1993 AJ 105, 2127
Cohen, M. 1993 AJ 105, 1860
Gilmore, G. 1984 MNRAS, 207, 223

Gilmore, G. and Reid, I. N. 1983 MNRAS 202, 1025

Kent, S. M., Dame, T. M. and Fazio, G. 1991 ApJ 378, 131

McCuskey, S. W. 1965 in Galactic Structure, A. Blaauw and M. Schmidt, eds., Univ. of Chicago Press, p. 1

McCuskey, S. W. 1966 Vistas in Astronomy, 7, 141

Mohan, V., Bijaoui, A., Crézé, M. and Robin, A. C. 1988 A&AS 73, 85

Philip, A. G. D. 1966 ApJS 12, 391

Robin, A. and Crézé, M. 1986a A&A 157, 71

Robin, A. and Crézé, M. 1986b A&AS 64, 53

Robin, A., Crézé, M. and Mohan, V. 1992a A&A 265, 32

Robin, A., Crézé, M. and Mohan, V. 1992b ApJ 400, L25

Smriglio, F., Nandy, K., Boyle, R.P., Dasgupta, A. K., Straižys, V. and Janulis, R. 1991 A&AS 88, 87

Straižys, V. 1973 Bull. Vilnius Obs., No. 36, 3

Straižys, V. 1977 Multicolor Stellar Photometry, Mokslas Publ. House, Vilnius, Lithuania

Straižys, V. 1992 Multicolor Stellar Photometry, Pachart Publ. House, Tucson, Arizona

Upgren, A. R. 1963 AJ 68, 475

Wainscoat, R. J., Cohen, M., Volk, K., Walker, H. J. and Schwartz, D. E. 1992 ApJS 83, 111

Wielen, R. 1974 Highlights of Astronomy, 3, 395

DISCUSSION

PHILIP: We have discussed the making of filters for Vilnius System CCD Photometry. Perhaps you could describe to the audience how you plan to provide Vilnius filters?

STRAIZYS: The filters of the Vilnius system are of two varieties: glass filters and interference filters. We make both types of filters in our Institute. The usual size of the filters for CCD photometry is 60 mm in diameter, the transmission at maximum is about 60-80%. Our optician can make the interference filters of larger size and for other wavelengths. We accept also orders from other institutions.

LEE: Do you plan to compare your system to existing photometric systems? If not, then how do you compare your conclusion to similar work done by other researchers using different photometric systems?

STRAIZYS: It is very important to compare the new results with other available information. However, our project will give MK spectral types and absolute magnitudes of stars down to 18-21st mag. For such faint stars usually no other classification will be available. Thus, a comparison will be possible only for brighter stars and not in all areas.

RATNATUNGA: With what model are you planning to use to compare your very useful star counts on the galactic plane? You should not extrapolate the Bahcall and Soneira Model, as that software is inappropriate to predict starcounts on the galactic plane.

STRAIZYS: It is important to compare our surface densities of stars with the predictions by the Bahcall and Soneira model. If necessary then new software will be made. Also, we shall compare our results with the models proposed by A. Robin and M. Crézé (A&A 157, 71 1986, A&AS 73, 85 1988, A&A 265, 32 1992) and by M. Cohen and his collaborators (ApJS 83, 111, 1992 and AJ 105, 1860, 1993).

A VERSATILE NEW MODEL FOR THE POINT SOURCE SKY

Martin Cohen

Jamieson Science & Engineering, Inc.
and Radio Astronomy Laboratory, University of California, Berkeley

ABSTRACT. I report further developments of the Wainscoat et al. (1992, hereafter WCVWS) model originally created for the point source infrared sky. This is now extended to handle the prediction of cumulative or differential source counts, and integrated surface brightness of the sky due to smeared point sources, in any galactic direction, for any infrared filter with passband within the range 2.0 - 35.0 μm. The already detailed and realistic representation of the Galaxy (disk, spiral arms and local spur, molecular ring, bulge, spheroid) and the extragalactic sky has been improved, guided by CO surveys of local molecular clouds. The newest version of the model is very well-validated by IRAS source counts. WCVWS demonstrated the validity of their model at V; it operates well at B. A major new aspect is the extension of the same model down to the FUV. I am now comparing predicted and observed FUV source counts from the Apollo 16 "S201" experiment (1400 Å) and the TD1 satellite (1565 Å). The FAUST passband has been incorporated in readiness for comparison with this experiment's deep 1660 Å images.

1. INTRODUCTION

The basic model (SKY) is described by WCVWS. It was originally required to provide predictions for mid-infrared point source counts. However, we felt it important *not* to design a model solely for the infrared but rather one that stressed geometric realism and accuracy of physical attributes of stars and non-stellar objects over a very broad range of wavelengths. SKY's wavelength capabilities are of two kinds. It operates with a number of broadband hardwired filters (standard Johnson B, V, K; IRAS [12] and [25]; and the Japanese narrowband balloon bandpass at 2.4 μm). It can also handle customized filters because its code draws upon an embedded complete library of continuous infrared spectra for all categories of source. WCVWS provided a library based on the IRAS Low Resolution Spectrometer (λ 7.7 - 22.7 μm). Cohen (1993) extended this spectrally continuous library to the 2.0 - 35.0 μm range. This provides the flexibility to make predictions for the present generation of infrared satellites (e.g. COBE/DIRBE) and the next (e.g. ISO).

2. GEOMETRY AND SOURCE TABLE

SKY incorporates five fully detailed galactic components: an exponential disk; a Bahcall-type bulge; a de Vaucouleurs halo; a four-armed spiral and the local spurs; and a circular molecular ring. Each geometrical element may contain up to 87 categories of source based upon detailed analyses of the content of the IRAS sky (Walker et al. 1989). These comprise 33 "normal" stellar types; 42 types of AGB star, both oxygen- and carbon-rich; six types of object that are distinct from others only in their mid-infrared luminosity; and six types of exotica

Workshop on Databases for Galactic Structure
A. G. Davis Philip, Bernard Hauck and Arthur
R. Upgren, eds. p. 131 - 136 © 1993 L. Davis Press

Martin Cohen
Radioastronomy Lab.
601 Campbell Hall
University of California
Berkeley, California 94720

including H II regions, planetaries and reflection nebulae. Every category of source has its own set of absolute magnitudes in the hardwired passbands; the dispersion in M_{12}; its own individual scale height and volume density in the local solar neighborhood. In addition, some sources may be absent from some geometrical components (the arms and ring are made rich in high-mass stars; the halo deficient in these). SKY also accommodates external galaxies, through their IRAS luminosity functions, but only for wavelengths beyond five μm.

These many attributes, all drawn from the copious astronomical literature, are entirely equivalent to the customary V and K luminosity functions. But our approach facilitates augmentation of the content of our model sky by adding other populations, or updating of physical parameters as these become better known from independent studies.

3. OUTPUT CAPABILITIES

SKY delivers either a cumulative or differential source count in any hardwired or customized filter (lying entirely between 2.0 and 35.0 μm) and can be compared in a single plot directly with observations. Inputs are either a single galactic direction, in which case it will predict for a 1 deg^2 patch centered on this (l,b), or a finite area (with chosen grid spacing in longitude and latitude), when the ouput will be the total source count in this area. It also yields the total integrated sky brightness in any filter, by smearing all the point sources expected in the region. Surface brightness information can be further broken down into the differential contributions of any of the individual geometrical elements (disk, halo, etc.). Abscissae can be magnitude or in-band flux, as desired. Color-color histograms are also implemented but this aspect represents work still in progress.

4. COMPARISON WITH THE MID-INFRARED (IRAS) SKY

WCVWS compared the performance of SKY with observed IRAS point source counts at 12 and 25 μm across the entire sky. They sampled 238 separate galactic zones, finding $> 80\%$ to have essentially perfect predictions at each wavelength, and some in which discrepancies were found. Nowhere did these discrepancies exceed a factor of three. This fidelity is quite remarkable because all the modeling in this first version of SKY was purely analytically expressed.

I have since improved the representation of the local spurs and enhanced the population densities in disk and arms, where the line-of-sight lies within the local spurs and intersects molecular clouds as mapped by Dame et al. (1987). The 12 μm skywide counts are now predicted almost perfectly, with only five zones discrepant by a factor as large as 1.5 - 2. The 25 μm sky is also improved; a few problems remain near the galactic plane and towards the Orion OB association.

5. TWO MICRON ABILITIES

Cohen (1993) compares SKY with K counts in the galactic plane by Eaton, Adams and Giles (1984) and Kawara et al. (1982). SKY performs well except towards l= 20 and 30°. In Baade's Window (cf. Ruelas-Mayorga and Teague 1992) SKY also does well (Fig. 1). These discrepancies towards 30° longitude, also apparent in recent comparisons with the TMGS data (see this volume), are such that SKY's predictions fall a factor of two to three below observations. Using the enhanced local spur version I do predict counts *above* all these observations. However, I have not yet incorporated the extra extinction that surely accompanies the enhanced stellar populations in these local molecular clouds. The match to IRAS at 12/25 μm examines the correctness of the stellar populations but these long wavelengths do not test the extinction. Therefore, K data are ideal for improving our representation of extinction with distance and galactic direction.

6. OPTICAL PERFORMANCE

WCVWS demonstrated the validity of their model at V, comparing it with data for the NGP assembled by Bahcall and Soneira (1984). It performs well at B as illustrated here by Terndrup's (1988) differential B counts for Baade's Window (Fig. 2). His counts fall away from SKY's predictions essentially where he estimates his incompleteness sets in.

7. FAR-ULTRAVIOLET CAPABILITIES

As part of a long-term development program for SKY, I am now embarked on an extension into the far-ultraviolet. So far I have hardwired 1400, 1565 and 1660 Å bands to handle the S201 Apollo 16 dataset (Page, Carruthers and Heckathorn 1982), the S2/68 experiment on the TD1 satellite (Thompson et al. 1978), and the Shuttleborne FAUST (Bowyer et al. 1993). Spectra were obtained from the IUE databases (Heck et al. 1984, Wu et al. 1991), and from Kurucz's (1991) library of synthetic spectra. I used these to obtain UV color indices, [UV]-[V]. So far I have included only the 33 categories of normal star, treating all the AGB stars with their dusty circumstellar shells as contributing nothing to the FUV. Likewise I have not yet included the exotica, whose UV spectra would often be dominated by emission lines. In spite of these simplifications, performance is very good at high and intermediate latitudes (Fig. 3). In the plane it is quite good, but I do not regard these predictions as meaningful until I improve the extinction model in the local clouds.

8. TYPICAL APPLICATIONS

With new near-infrared sky surveys proposed and under way it is valuable for the design of survey strategies to assess those regions of the sky in which confusion due to high source density may render it difficult, if not impossible, to complete the surveys. Confusion is also important to the design of experiments from space, such as ISO. SKY can be used in its cutomized mode to determine the longest meaningful exposures in and near the galactic plane, to select appropriate filters to achieve observational goals, and to quantify the degradation of limiting performance as a function of line-of-sight direction (e.g. Fig. 4).

Matsumoto, Akiba and Murakami (1988) claimed detection of a cosmological background at two μm from their rocket sky measurements. To recognize such a background one must first subtract the scattering from zodiacal dust, then remove the expected contribution of foreground galactic stars. These authors used a very simple model to estimate the stellar foreground surface brightness. Even if we assume that their zodiacal contribution was correctly assessed, SKY predicts 2.4 times as much smeared starlight as Matsumoto et al.(1988) effectively eliminating the alleged cosmological contribution. To separate galactic foreground starlight from cosmological

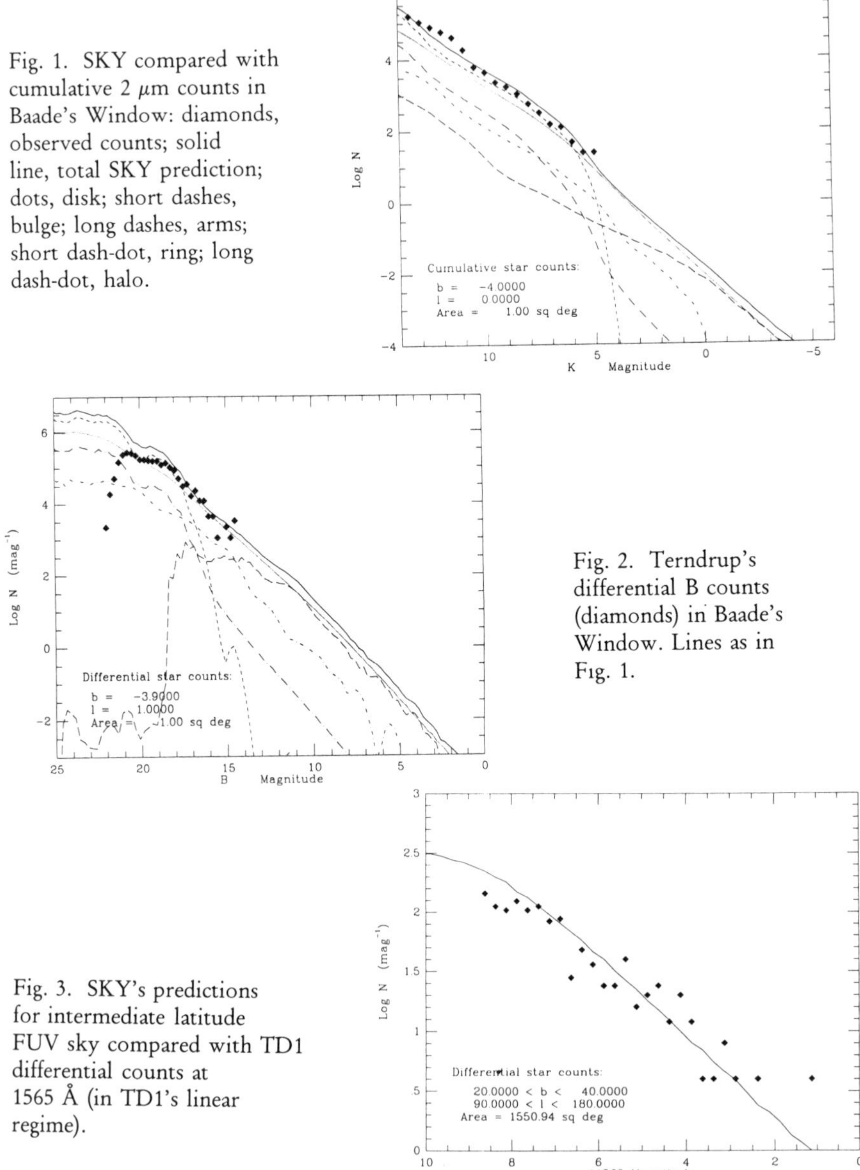

Fig. 1. SKY compared with cumulative 2 μm counts in Baade's Window: diamonds, observed counts; solid line, total SKY prediction; dots, disk; short dashes, bulge; long dashes, arms; short dash-dot, ring; long dash-dot, halo.

Fig. 2. Terndrup's differential B counts (diamonds) in Baade's Window. Lines as in Fig. 1.

Fig. 3. SKY's predictions for intermediate latitude FUV sky compared with TD1 differential counts at 1565 Å (in TD1's linear regime).

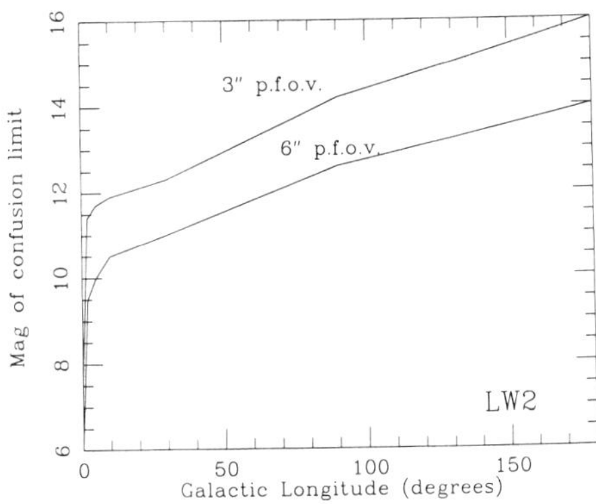

Fig. 4. Degradation of limiting magnitude due to high source density confusion in the galactic plane, calculated for ISOCAM's filter LW2 (5.0-8.5 μm). Two curves are plotted, for 3" and 6" pixel fields of view.

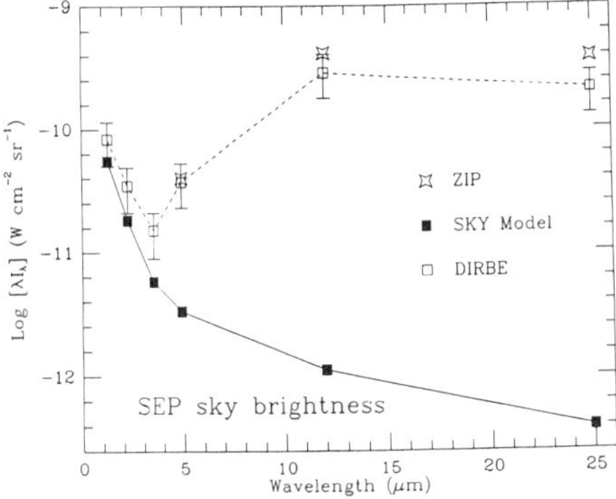

Fig. 5. SKY Predictions for surface brightness at the South Ecliptic Pole customized for DIRBE bands 1-6. The ZIP experiment essentially shows the shape and level of zodiacal emission.

or even diffuse galactic background as measured by COBE/DIRBE will clearly require an accurate and detailed model such as SKY (Fig. 5).

9. CONCLUSIONS

This brief overview is intended to highlight some of the powerful capabilities of SKY from the far-ultraviolet to the mid-infrared. I look forward to making comparisons with the new sky surveys discussed here, and to using improved parameters (scale heights and lengths, population densities, and especially M_V determined from the new databases described at this workshop.

ACKNOWLEDGMENTS

Past and future development of this model is supported by Phillips Laboratory through Dr. Stephan D. Price.

REFERENCES

Bahcall, J. N. and Soneira, R. M. 1984 ApJS 55, 67

Bowyer, S., Sasseen, T. P., Lampton, M. and Wu, X. 1993, preprint
Cohen, M. 1993 AJ 105, 1860
Dame, T. M., Ungerechts, H., Cohen, R. S., de Geus, E. J., Grenier, I. A.,
    May, J., Murphy, D. C., Nyman, L.-A. and Thaddeus, P. 1987 ApJ 327, 706
Eaton, N., Adams, D. J. and Giles, A. B. 1984 MNRAS 208, 241
Heck, A., Egret, D., Jaschek, M. and Jaschek, C. 1984 ESA SP-1052
Kawara, K., Kozasa, T., Sato, S., Kobayashi, Y., Okuda, H. and Jugaku, J.
    1982, PASJ 34, 389
Kurucz, R. L. 1991 NATO ASI Ser.C, Math. Phys. Sci., 341, 441
Matsumoto, T., Akiba, M. and Murakami, H. 1988 ApJ 332, 575
Page, T., Carruthers, G. R. and Heckathorn, H. M. 1982 NRL Rep. 8487
Ruelas-Mayorga, R. A. and Teague, P. F. 1992 A&AS 93, 61
Terndrup, D. M. 1988 AJ 96, 884
Thompson, G. I., Nandy, K., Jamar, C., Monfils, A., Houziaux, L., Carnochan,
    D. J. and Wilson, R. 1978 Catalogue of Stellar UV Fluxes, SRC, UK
Wainscoat, R. J., Cohen, M., Volk, K., Walker, H. J. and Schwartz, D.
    E. 1992 ApJS 83, 111 [WCVWS].
Walker, H. J., Cohen, M., Wainscoat, R. J. and Schwartz, D. E. 1989 AJ 98, 2163
Wu, C.-C., Crenshaw, D. M., Blackwell, J. H., Wilson-Diaz, D., Schiffer III,
    F. H., Burstein, D., Fanelli, M. N. and O'Connell, R. W. 1991 IUE UV
    Spectral Atlas: Addendum I, Newsletter 43

DISCUSSION

RATNATUNGA: The definition of the Galaxy by a set of components is very similar to that used in the IAS-Galaxy Model. However, inherited from the former BAS Galaxy Model is the representation of scale height as a function of absolute magnitude when it is a function of age. This is critical for kinematics. I would have expected the difference to be also important for counts of late-type stars on the galactic plane.

COHEN: It is very easy for me to add populations or to split existing ones into "young" and "old". For example, all it takes is scale height, colors and local solar neighborhood density. I don't think you would see any difference in improved density. I detect the K dwarf stars, for example, but it would be easy enough to check.

CREZE: Did you try to compare your model prediction in the disk with H I distribution?

COHEN: Yes, because in some zones we found no H I and there the model predictions exceed IRAS counts. So now I have reduced the disk contributions in those directions based on H I deficiency.

VAN ALTENA: Do you intend to incorporate kinematics into your model?

COHEN: I have my work cut out for the next few years extending the spectral library in the model from 2 - 35 μm! I've not considered adding kinematics, until this meeting.

TYCHO REFERENCE CATALOGUE (TRC): A CATALOGUE OF POSITIONS AND PROPER MOTIONS OF ONE MILLION STARS

Siegfried Roeser

Astronomisches Rechen-Institut

Erik Hoeg

Copenhagen University Observatory

ABSTRACT: The catalogue described here will be an extension of the Hipparcos system to higher star densities and fainter magnitudes. It will be produced by combining the Tycho catalogue with the Astrographic Catalogue (AC). The AC contains 10 million measurements of some 4 million stars. We plan to perform a new reduction of the AC using Hipparcos as reference. With the accuracy of the AC lying between 0.2 and 0.4 arcsec per measured coordinate depending on observatory zone, an average accuracy of 2.5 mas per year for the proper motions of 1 million stars is a realistic estimate. In this paper we shall call this catalogue TRC (Tycho Reference Catalogue). It will be produced as a collaboration between scientists in Copenhagen, Lund, Moscow and Heidelberg.

1. HIPPARCOS AND TYCHO

The HIPPARCOS Catalogue will supply a system of positions and proper motions of some 120,000 stars on the whole sky. The expected rms error of the proper motions will be about 2 mas/year. The Tycho experiment on Hipparcos will give the positions of 1 million stars to V = 12 mag with an accuracy of 30 mas (Hoeg et al. 1992). Proper motions of high accuracy are not expected from the Tycho experiment. They can only be derived by combining the Tycho astrometric measurements with suitable ground-based observations. The Astrographic Catalogue is the only source for such, having early epoch and appropriate accuracy.

2. REDUCTION OF THE ASTROGRAPHIC CATALOGUE

In recent years, the printed volumes of the Astrographic Catalogue have been transferred into machine-readable form at the Sternberg Astronomical Institute, Moscow (Nesterov et al. 1990). A careful reduction of the Astrographic catalogue is essential for good-quality proper motions. It will be the most critical task in the project. A first reduction of the AC has been carried out in the course of the production of the PPM catalogue (Roeser and Bastian 1991). In the northern hemisphere a systematically corrected version (Roeser 1990) of the AGK3 was used as reference catalogue. In the southern hemisphere an improved version - by inclusion of the CPC2 and the final FOKAT catalogue - of the Preliminary PPM South (Bastian et al. 1991) served for the reduction. Some statistics resulting from this reduction are presented in Table

Workshop on Databases for Galactic Structure
A. G. Davis Philip, Bernard Hauck and Arthur
R. Upgren, eds. p. 137 - 143 © 1993 L. Davis Press

137

Siegried Roeser
Astronomisches Rechen-Institut
Moenchhofstr. 12 - 14
D-6900 Heidelberg 1, Germany

1. The accuracy of star positions in the keypunched printed AC was determined by forming the differences of the measurements on overlapping plates. Column 4 in Table 1 gives the average rms error of a coordinate per single plate for the different AC zones. We see that there is almost a factor of two in accuracy between the best and the worst zones. The implications will be discussed in the next section.

The Hipparcos catalogue is, of course, the best candidate as reference catalogue for the AC. Only Hipparcos supplies a coordinate system at the epoch of the AC, which is free from zonal systematic errors, magnitude and color equations. The disadvantage is the low star density which gives an average of 15 stars per plate, and the very low number of faint (V > 10 mag) stars.

The first problem can be overcome by performing a plate overlap solution. Choosing, for example, the 12 surrounding plates for a common reduction of the central plate increases the number of reference stars to about 100. However, from Table 1 we find a large scatter in epoch from plate to plate in many zones of the AC. Therefore a simple plate overlap solution using all the overlapping AC stars is not possible, because their proper motions are unknown. That is why we shall select only stars present in the Tycho catalogue for the overlap, because in this case proper motions can reasonably be introduced as additional unknowns into the overlap equations.

The second problem of magnitude and color equations in the AC reduction is illustrated in Fig. 1. Fig. 1a presents the distribution of V-magnitudes in the Hipparcos catalogue, Fig. 1b the magnitude distribution in the southern Hyderabad zone as an example of the magnitude distribution in the AC. Noting that the AC magnitudes are essentially B and uncertain by 0.5 mag, we realize, that between 10.5 and 12 mag in V the number of Hipparcos stars is very small, whereas the bulk of the AC stars is falling into this magnitude range. Due to the small number of Hipparcos stars, we can only study a global magnitude equation within the individual AC-zones. Assuming a uniform distribution we find about 80 Hipparcos stars per zone in the V-magnitude range from 11.0 to 11.5 mag. Since there is no other reference catalogue - free of a magnitude equation - with higher star density at the faint magnitude end and at the epoch of the AC, we rely on Hipparcos hoping for successful determination of the magnitude equation in the AC-zones.

Due to the large epoch difference between Tycho and the AC, it is desirable to have another catalogue at an intermediate epoch for checking problem cases and stars with large proper motions. The only source serving this purpose is, at the moment, the Guide Star Catalogue (GSC) (Lasker et al. 1990). Combining the GSC with the AC we will produce, as a by-product of the TRC, a catalogue of 4 million stars, whose properties are discussed below. Note that the TRC will be an ideal catalogue for a new reduction of the GSC.

3. EXPECTED PROPERTIES OF THE TRC CATALOGUE

After the reduction of the AC and the determination of proper motions, we expect rms-errors of a proper motion component of the TRC as shown in the last column of Table 1. The rms-errors of the individual proper motions vary from one AC-zone to the other according to the accuracy of the AC positions and the epochs of observation. The best performance with

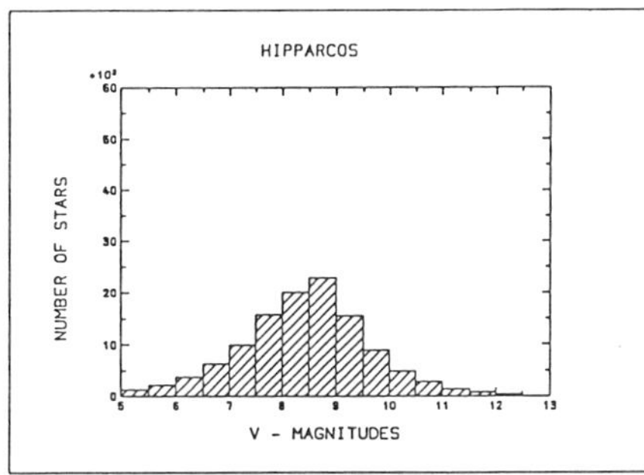

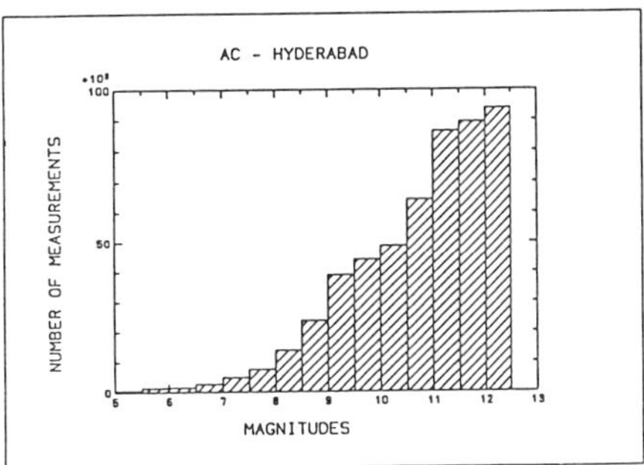

Fig. 1a. Histogram of the magnitudes of the stars in the Hipparcos Input Catalogue.

Fig. 1b. Histogram of the magnitudes of the stars in the southern Hyderabad zone of the Astrographic Catalogue.

formal individual accuracies equal or better than 2 mas/year will be reached in the Paris and Helsingfors zones. The worst results are expected in the regions that cover the uncompleted Potsdam zone. Due to the late mean epoch and the low accuracy of the measurements, stars in the Uccle zone between +33 and +36 degrees declination will obtain an accuracy of 6.5 mas/year, only. If the Uccle zone were to be remeasured, and an accuracy of 200 mas could be reached with modern automatic measuring engines - provided that the Uccle plates were still in good shape - the resulting accuracy of the Uccle proper motions would be 3.3 mas/year, and therefore quite satisfactory. The average rms-error for a proper motion component in the TRC is 2.5 mas/year, its mean epoch is 1989.3 and the rms-error of a positional coordinate at mean epoch 29 mas. Of course both values are close to the the Tycho values.

In Table 2 the performance of the TRC catalogue is compared with the PPM catalogue and the Hipparcos catalogue. As can be seen from this table, the TRC will surpass the PPM both in the number of stars by a factor of 2.5 and in the accuracy of proper motions by a factor of 1.7 in the north and 1.2 in the south. The TRC will almost reach the individual accuracy of the Hipparcos proper motions, and it has about 8 times as many stars as Hipparcos. The most important, but so far unanswered question is the problem of systematic accuracy of the proper motions. For comparison we have introduced in Table 2 the GSC-AC, where we adopted an rms-error per coordinate of 0.5 arcsec (Russell et al. 1990) at a mean epoch of 1980.

With its high spatial density of 25 stars per square degree, the TRC will be a very valuable reference catalogue for photographic astrometry. Its rms accuracy of a coordinate in the year 2020, for instance, will be about 85 mas, compared with 58 mas and a density of 2.5 stars per square degree in Hipparcos. The comparison between the PPM, Hipparcos and the TRC is shown in Fig. 2. As a side remark we note, that it will be unnecessary to observe any TRC star with meridian circles within the next decade. Unless replaced by an even better catalogue, the TRC will be the reference catalogue for photographic astrometry in the first part of the 21st century.

Finally, we ask about the importance of the TRC catalogue for kinematics in our galaxy.

TABLE 1
Zones of the Astrographic Catalogue

Zone	Mean Epoch	Sig (M.Ep.)	Sig (Pos.) arcsec	P.M mas/year
Greenwich	1897.3	2.8	0.28	2.1
Vatican	1908.5	5.5	0.42	3.6
Catania	1903.1	3.6	0.36	2.9
Helsingfors	1895.0	2.2	0.26	1.9
Potsdam	1895.9	2.1	0.25	1.9
Hyderabad N	1931.1	2.1	0.36	4.3
Uccle	1944.7	3.5	0.42	6.5
Oxford II	1933.8	2.0	0.31	3.9
Oxford I	1900.7	4.3	0.32	2.5
Paris	1895.7	3.4	0.23	1.7
Bordeaux	1905.6	5.0	0.26	2.2
Toulouse	1910.2	9.9	0.25	2.2
Algiers	1904.5	7.1	0.25	2.1
San Fernando	1897.2	6.0	0.31	2.4
Tacubaya	1904.5	2.8	0.30	2.5
Hyderabad S	1919.1	2.7	0.38	3.8
Cordoba	1912.5	3.2	0.35	23
Perth	1911.6	3.3	0.37	33
Cape	1902.8	2.5	0.31	52
Sydney	1907.7	3.6	0.43	63
Melbourne	1898.1	8.1	0.30	2.4

The columns have the following meaning: Name of observatory, mean epoch of the observations, dispersion of epochs, rms-error of a coordinate, estimated mean error of a proper motion in the TRC (assuming 30 mas for Tycho at epoch 1991).

TABLE 2
Properties of Catalogues

Name	N	Mean epoch	Sig (Pos.) mas	Sig(P.M) mas/year
PPM North	180,000	1931	91	4.2
PPM South	200,000	1962	72	3.0
Hipparcos	120,000	1991	02	2.0
TRC	1,000,000	1989	29	2.5
GSC-AC	4,000,000	1916	200	6.9

The columns give the name of the catalogue, the number of stars, the mean epoch, the mean error of a position coordinate at mean epoch and the mean error of a proper motion coordinate.

To answer this question, we will compare the number of stars with significant proper motions in various catalogues. By a significant proper motion we understand a total proper motion which exeeds three times its estimated standard deviation. As we do not know, a priori, the distributions of total proper motions in those catalogues, which do not yet exist, we have taken the distribution of the PPM proper motions as representative. Fig. 3 shows this distribution. It has its maximum at proper motions of 11 mas/year, its median at 20 mas/year, and its mean at 30 mas/year. The distribution for Hipparcos will not differ significantly from that of the PPM, because the set of Hipparcos stars essentially is a subset of the set of the PPM stars; more than 90 per cent of the Hipparcos stars are in the PPM. For the TRC, and even more for the GSC-AC, the distribution will be shifted towards smaller proper motions. Table 3 compares the different catalogues with respect to their contents of stars having significant proper motions. The columns in this table have the following meanings: name of the catalogue, lower bound of significant proper motions, number and percentage of stars having significant proper motions. The final column is a calculated secular parallax (averaged over the sphere) corresponding to the value of the proper motion in column 3. Of course, the Hipparcos catalogue will have the highest percentage of stars having significant proper motions, followed by the TRC and the PPM-South. Even in these three catalogues only proper motions beyond the peak of the proper motion distribution are significant, in the PPM-North significance starts

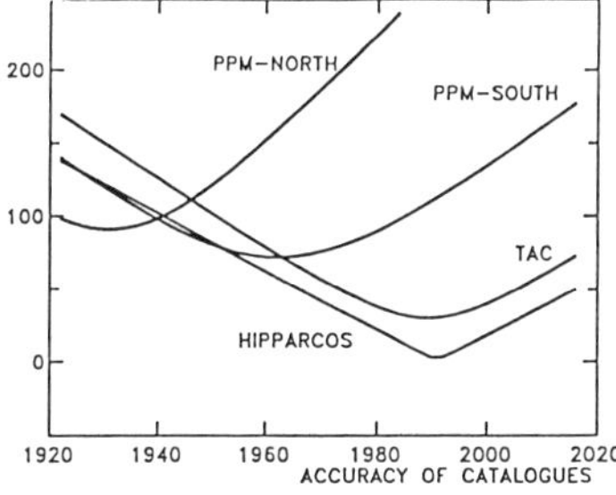

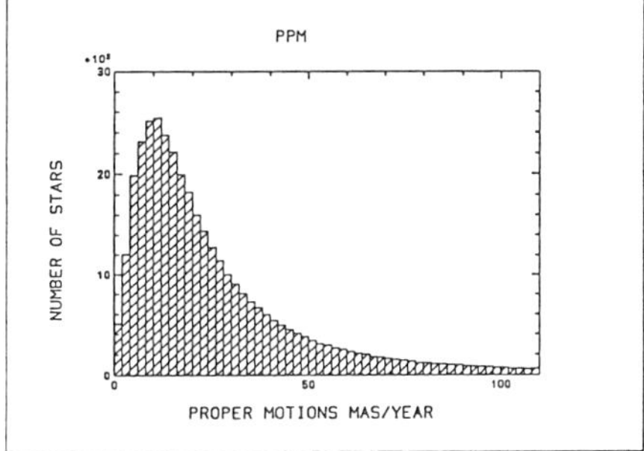

Fig. 2. RMS-errors of a coordinate (R.A. or Decl.) as a function of time in different catalogues.

Fig. 3. Distribution of the total proper motions in the PPM.

at the median, whereas in the GSC-AC only high-velocity stars in the tail of the distribution have significant proper motions. In total number, Hipparcos has 100,000 significant proper motion stars. The TRC contains essentially all Hipparcos stars, therefore it adds 670,000 stars with this property. Although the criterium of "significant proper motions" is by no means exhaustive, it points to the importance of catalogues for kinematics in our galaxy. We see, that the TRC will be the most important catalogue in this respect. From the lower bound of significant proper motions we can calculate an upper bound for a secular distance. This distance marks the volume within which we can expect significant proper motions in the different

TABLE 3

Significant proper motions

Name	3*Sigma mas/y	Number of sign. stars	Percentage of sign. st.	Distance pc
PPM North	18	100,000	55	190
PPM South	12	140,000	70	275
Hipparcos	8.4	100,000	84	390
TRC	10.5	770,000 (670,000)	77	320
GSC-AC	30	1,350,000 (580,000)	34	110

The numbers in brackets give the additional significant proper motions with respect to the catalogue in the row above. (See text for further explanation).

catalogues. Again, Hipparcos fills the largest volume but with the lowest number of stars. The TRC, in almost the same volume, supplies six times as many stars as Hipparcos.

Formally, the GSC-AC will contribute some 600,000 significant proper motion stars. But, this conclusion is very doubtful. In the GSC-AC only stars with relatively high proper motions will be significant. The secular distances for these stars should be smaller than 110 pc. But stars in the GSC-AC are typically fainter than in the other catalogues and are, on the average, farther away. This discrepancy shows that the assumption, that the proper motion distribution in the GSC-AC has the same shape as in the PPM, is not correct.

REFERENCES

Bastian, U., Roeser, S., Nesterov, V. V., Polozhentsev, D. D., Potter, Kh.I., Wielen, R., Yagudin, L.I. and Yatskiv, Ya. S. 1991 A&AS 87, 159

Hoeg, E., Bastian, U., Egret, D. et al. 1992 A&A 258, 177

Lasker, B. M., Sturch, C. R., McLean, B. J., Shara, M. M., Russell, J. L. 1989 AJ 99, 2019

Nesterov, V. V., Kislyuk, V. S. and Potter, Kh. I. 1990 in IAU Symposium 141, Inertial Coordinate System on the Sky, J. H. Lieske and V. K. Abalakin, eds., Kluwer Academic Press, p. 482

Roeser, S. 1990 in IAU Symposium 141, Coordinate System on the Sky, J. H. Lieske and V. K. Abalakin, eds., Kluwer Academic Press, p. 469

Roeser, S. and Bastian, U. 1991 PPM Star Catalogue. Astronomisches Rechen-Institut, Heidelberg, Spektrum Akademischer Verlag.

Russell, J. L., Lasker, B. M., McLean, B. J., Sturch C. R., Jenkner, H. 1989 AJ, 99, 2059

DISCUSSION

GRIFFIN: How valuable do you consider it would be to have measurements of the radial velocities too many of those stars?

ROESER: They are very valuable additions.

GRIFFIN: A few years ago a proposal was made in the UK to purchase and operate two 1-meter automatic telescopes and Coravel-type machines, one in the north and one in the south, but it was discussed by the peer-astronomer committee as of no significant interest.

URBAN: Tycho results will be extremely valuable. I believe it will contain three times the stars than originally planned with little or no loss of accuracy. The distribution of Hipparcos data is well known. What are the plans of distributing the Tycho data?

ROESER: I can give here only my personal impression, I cannot give an official statement. I think the Tycho result will be available in 1997, the TRC catalog may appear in 1998.

LU: Do you have plate overlapping solution for all AC catalogs? Twenty years ago I worked on the Melbourne Zone AC catalogs and I am aware that each AC catalog plate has its own plate solution. Did you have any comparison between current plate overlap solution with the original plate solution?

ROESER: We have plate overlap solutions for the southern hemisphere. I am sorry to say that we had no time to compare the solutions using the original plate constants with our new reductions.

U. S. NAVAL OBSERVATORY'S ASTROGRAPHIC CATALOG PROJECT

Sean E. Urban

U. S. Naval Observatory

1. INTRODUCTION

The current status of proper motion databases is quite good for stars brighter than magnitude 10.5. The Smithsonian Astrophysical Observatory Catalog (SAOC), which has been a source of positions and proper motions since the mid-sixties, has now been superseded by the Astrographic Catalog Reference Stars (ACRS). Both catalogs contain data on stars down to magnitude 10.5 and are fairly uniformly distributed over the celestial sphere. For positions and motions of stars fainter than 10.5, the scenario is bleak. There is a lack of available positional data for faint stars to be used to calculate proper motions. The Astrographic Catalog Project (AC project) underway at the U.S. Naval Observatory (USNO) will improve this situation. The AC project is designed to make new reductions of the Astrographic Catalogue plates. These photographic plates, with epochs about 1905, are an excellent source of first epoch positions, that, when combined with recent positional data, will lead to highly accurate proper motions for stars down to magnitude 12.5.

2. CURRENT STATE OF LARGE DATABASES OF PROPER MOTIONS

Two of the better known proper motion databases are summarized below.

2.1 Smithsonian Astrophysical Observatory Catalog

The Smithsonian Astrophysical Observatory Catalog (SAOC, 1966) has probably been the best known and most widely used catalog of positions and proper motions. It was originally constructed for satellite tracking, and its goal of having a mean error of position no larger than 1"0 at epoch 1970 was met. With more than 250,000 stars, it was the largest database for proper motions at the time of its release in the mid-1960's. At the current epoch, however, the SAOC is unsatisfactory for most astrometric applications. The positions and proper motions have large individual and systematic errors. The original goal of the SAOC was met, but the need for highly precise astrometric data makes the SAOC obsolete for many users.

2.2 The Astrographic Catalog Reference Stars

The Astrographic Catalog Reference Stars (ACRS) (Corbin and Urban, 1991) was compiled at the USNO for the purpose of making new reductions of the Astrographic Catalogue plates. The ACRS is a pole-to-pole catalog containing 320,211 star positions and proper motions. In all, 1.6 million positions from 167 catalogs were used in compiling the ACRS. Extreme care

Workshop on Databases for Galactic Structure
A. G. Davis Philip, Bernard Hauck and Arthur
R. Upgren, eds. p. 145 - 151 © 1993 L. Davis Press

Sean E. Urban
U. S. Naval Observatory
3450 Massachusetts Ave.
Washington D. C.
20392

was taken to ensure that the ACRS would be systematically consistent. Each of the 167 individual input catalogs was reduced to the FK4 system by use of the International Reference Stars (IRS) (Corbin 1978, Corbin and Urban 1990). With an average proper motion mean error of 0"47/century and an average epoch of 1949.2, it is currently the most accurate database of its kind. Fig. 1 shows the relative positional accuracy of the ACRS and the SAOC for epoch 1993. For comparison purposes, positional accuracy of the FK5 is also included. The ACRS should be the choice of astronomers who need high quality position and proper motion data for stars brighter than 10.5.

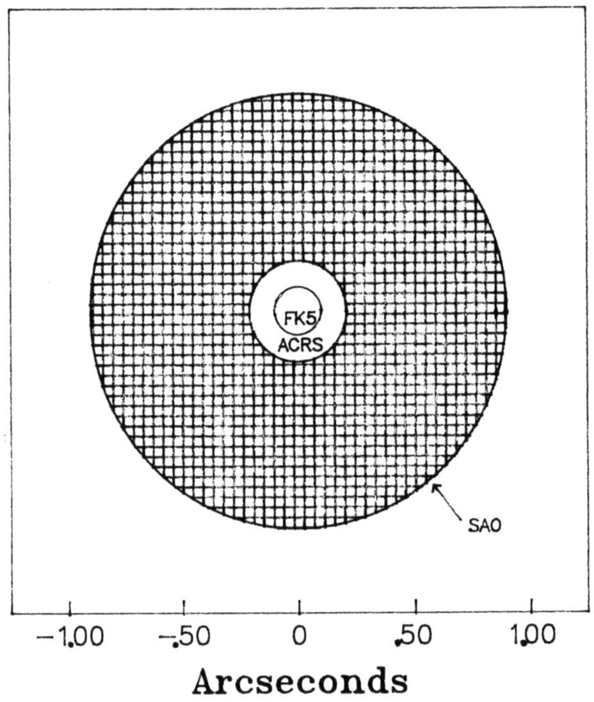

Fig. 1. Relative positional accuracies at 1993 for the FK5 basic, the ACRS, and the SAO catalogs are shown.

Each of these catalogs is limited to about magnitude 10.5, with roughly the same star density, 6-8 stars/degree^2. What is desired is a much larger, fainter, accurate database of positions and proper motions. Unfortunately, such a database does not exist at this time. There is, however, a source of star positions that can be used to make a great stride toward an accurate proper motion database of faint stars; it is the Astrographic Catalogue.

3. THE ASTROGRAPHIC CATALOGUE

In the late 1800's, technological advances in telescope design and photography gave birth to the idea that all stars, down to a certain limiting magnitude, could be charted using photographs. Following an international congress held in 1887, a program to catalog the entire sky using photographic plates was established. In all, 20 observatories with sites worldwide participated. The result of this work, photographic positions of 5 to 10 million stars, is known as the Astrographic Catalogue (AC). Most of the AC plates were exposed between 1891 and 1920, giving the AC a mean epoch about 1905. The limiting magnitude, although it varies with each plate, is approximately 12.5. It is this early epoch and magnitude limit that makes the AC an ideal first-epoch catalog for proper motion databases. Fig. 2 shows the accuracy of proper motions that can be achieved when the AC positions are combined with a late-epoch catalog. The solid curve shows the mean error of proper motions using a positional accuracy of 0"15 for a 2nd epoch catalog and a conservative value of 0"30 for the AC. The dotted curve uses the value of 0"10 for a second-epoch catalog and a value of 0"20 for the AC. Since we are now approaching a 90-year difference from the AC epoch, it is quite possible to determine proper motions with an accuracy of 0"35/century for many of the AC's 5 to 10 million stars.

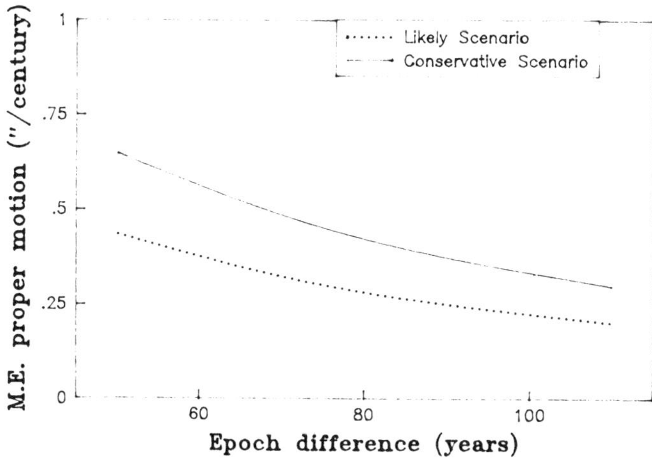

Fig. 2. Shown are the accuracies of two position proper motion as a function of the epoch difference of the positions. The dotted line represents the accuracy using positional mean errors for the two catalogs of 0"20 and 0"10. The solid line uses positional mean errors of 0"30 and 0"15.

Unfortunately the Astrographic Catalogue, as it currently exists, is not suitable for a first-epoch proper motion database. When the project was first established, participating observatories were given guidelines on the type of hardware to use and on the photographic techniques to be employed. However, the type of reduction methods and publication formats were, for the most part, left up to the individual observatories. As a result, the AC is very inhomogeneous, making it quite difficult to use. Another problem is that most zones of the AC do not contain positional data in spherical coordinates at all; instead, the star positions are given only in x, y plate measures and provisional plate constants. The reference stars used to determine those plate constants vary systematically with each zone, so the resulting positions contain serious systematic errors. Others have demonstrated that these errors can be greatly reduced by using a high quality reference star catalog. One more problem is that much of the AC was not in machine-readable form. Fortunately, the U.S. Naval Observatory is committed to untangling the inconsistencies found in the AC, making new reductions of its plates, and making all data available.

4. THE ASTROGRAPHIC CATALOG PROJECT OF THE U. S. NAVAL OBSERVATORY

The potential uses of the Astrographic Catalogue are recognized by the USNO and the USNO is committed to its re-reduction. Over the last several years, many of the printed AC volumes have been transferred to magnetic tape. The ACRS, a reference star catalog specifically designed be used with the AC plates, has been compiled. Plate reduction software is currently in place and reductions of individual plates are proceeding.

4.1 X, Y Data Transfer

For the last several years, the USNO has continued work, started in Strasbourg (Valbousquet 1977), of transferring the printed AC data onto magnetic tape. All x-y's are now in machine-readable form for the entire Southern Hemisphere and from the Equator to declination $+40°$. Work is continuing on the zones from $+40°$ to $+90°$ Checks are made to ensure that all x, y's on all plates are verified.

4.2 Astrographic Catalog Reference Stars

Although it has many uses, the ACRS (see Section 2) was first constructed to give a systematically consistent, dense set of reference stars to be used for new reductions of the AC plates. Prior to its release, no reference star catalog was suitable for the job (Corbin and Urban, 1988). As a result of the high density of the ACRS, each Astrographic Catalogue plate contains about 30 ACRS stars.

4.3 Plate Reductions

The U.S. Naval Observatory has implemented software for new reductions of the AC plates. Much of the software was originally developed at Hamburg Observatory (Zacharias et al. 1992) and later modified for computing systems at USNO. It is envisioned that a conventional plate adjustment will be made (that is, individual plates reduced separately) followed by a global block adjustment solution. The first band being reduced at USNO was photographed by the Royal Observatory at the Cape of Good Hope (declination range -51° to -41°). It is hoped that the Cape zone will be ready for distribution by the end of 1993, and three to four zones per year will follow.

As previously demonstrated in Fig. 2, the true value of the AC will be realized when proper motions for its stars can be established.

5. THE AC AS A FIRST EPOCH FOR FUTURE PROPER MOTION DATABASES

The Astrographic Catalogue contains positional data of 5 to 10 million stars at epoch 1905. These positions, in combination with newer catalogs, can yield outstanding proper motions. Currently there is no individual catalog that can be used to give proper motions of all of the AC's stars, but there are a few projects under way that will give late-epoch positions of a high percentage of them.

5.1 U.S. Naval Observatory Northern Astrograph Program

The USNO 8-inch twin astrograph was used to photograph the sky from +90° to -10° declination during the years 1977 to 1985. These plates have a limiting magnitude of 11.0 and are currently under reduction (Douglass 1993). This project will give second-epoch positions to approximately 600,000 AC stars.

5.2 Tycho Astrometric Catalogue

The Tycho mission is the lesser known half of the Hipparcos satellite mission. The original Tycho mission was designed to give a survey of approximately 300,000 stars down to magnitude 11 (Egret and Hoeg, 1988). The satellite was also designed to scan areas around each of the survey stars in an attempt to derive positions of stars fainter than 11th magnitude. This additional scan mode has made it likely that the satellite will determine positions for up to 1 million stars down to magnitude 12.5, with accuracies better than $0''\!.1$ (Hoeg et al. 1992). These late-epoch positions would lead to proper motions accurate to about $0''\!.3$/century for 10% of the AC stars.

5.3 New Measurements of the AGK2 Plates

The plates of the Second Astronomische Gesellschaft Catalog (AGK2) (Schorr and Kohlschutter, 1951-58), a photographic catalog of the Northern Hemisphere, were originally measured to a magnitude of 10.5. The photographic plates, however, have a limiting magnitude of approximately 12.0. These plates are in good condition and are stored at the the Hamburg Observatory. It is reasonable that positions resulting from a remeasurement of the AGK2 would have a mean error of about $0''\!.15$. With the AGK2 epoch at 1930, the resulting mean error of proper motions, when combined with the AC, would be about $1''\!.1$/century. The best use of a remeasurement of the AGK2 plates would be to combine the resulting positions with the AC and a late-epoch catalog to determine proper motions. In this scenario, the AGK2 would be a useful tool for strengthening the proper motions of the AC stars.

6. SUMMARY

The status of proper motions for stars brighter than magnitude 10.5 is quite good, the ACRS being the best catalog for most users' needs. For stars fainter than 10.5 the proper motion database situation is severely limited. The Astrographic Catalog project, currently underway at the U.S. Naval Observatory, will yield systematically consistent positions accurate to 0"2 to 0"3 for the 5 to 10 million stars in the AC. The result of these positions, in combination with newer observations, will be a database of proper motions with accuracies about 0"35/century for stars down to magnitude 12.5

REFERENCES

Corbin, T. E. 1978 in IAU Symposium 109, Modern Astrometry, F. V. Prochazka and R. H. Tucker, eds., University Observatory, Vienna, p. 505

Corbin, T. E. and Urban, S. E. 1988 in IAU Symposium 133, Mapping the Sky, S. Debarbat, J. A. Eddy, H. K. Eichhorn and A. R. Upgren, Kluwer Academic Pub., Dordrecht, p. 287

Corbin, T. E. and Urban, S. E. 1990 in IAU Symposium 141, Inertial Coordinate System on the Sky, J. Lieske and V. K. Abalakin, eds., Kluwer Academic Pub., Dordrecht, p. 433

Corbin, T. and Urban, S. 1991 Astrographic Catalog Reference Stars, NASA, NSSDC 91-10

Douglass, G. 1993, private communication

Egret, D. and Hoeg, E. 1988 in IAU Symposium 133, Mapping the Sky, S. Debarbat, J. A. Eddy, H. K. Eichhorn and A. R. Upgren, eds., Kluwer Academic Pub., Dordrecht, p. 265

Hoeg, E. et. al, 1992, Astron. Astrophys. 258, 177

Schorr, R. and Kohlschutter, A. 1951-1958 Zweiter Katalog der Astronomischen Gesellschaft fur das Aquinoktium 1950, v. 1-10, Hamburg-Bergedorf Sternwarte; v. 11-15 Dummler, Bonn

Smithsonian Astrophysical Observatory Star Catalog, SAO staff 1966 Smithsonian Astrophysical Observatory

Valbousquet, A. 1977 Bull. Inform. CDS, 13, p. 2

Zacharias, N., de Vegt, C., Nicholson, W. and Penston, M. J. 1992 A&A 254, 397

DISCUSSION

LEE: Are FK5 extension stars also not included in ACRS catalog?

URBAN: In truth, the basic FK5 and FK5 Supplemental stars (that is, stars with FK5 numbers less than 4000) are not in the ACRS catalog. The FK5 Faint Fundamental Extension stars (FK5 numbers 4000 and higher) are included in the ACRS.

CREZE: What is the expected accuracy of ACRS positions at the epoch of AC plates?

URBAN: Although it varies, the accuracy is generally 0.2 to 0.25 arc seconds.

CREZE: Has there been any comparison made between the ACRS and the PPM? Roeser made a comparison between global and local proper motions, but the error budgets that were presented here, although referring to global proper motions, are only based on local error considerations, could any of the speakers comment on that?

URBAN: Roeser actually made the comment that those who would like to compare the catalogs should do just that, but the catalogs (since input lists contained the same data) are correlated.

LASKER: The logic of accepting the original measure on a collection of 22,000 plates is obvious. But, are all of the plate measures usable? Do any plates need to be remeasured?

URBAN: With the accuracy of modern measuring machines, there is no question that some zones (if not all) would benefit from remeasurement. However some questions must be answered such as: Are the plates still in good condition? Are observatories holding the plates willing to give them up for a time? And who pays?

THE SOUTHERN PROPER MOTION PROGRAM: INPUT CATALOG PREPARATION

I. Platais, T. M. Girard, W. F. van Altena

Yale Observatory

C. E. Lopez

Felix Aguilar Observatory and Yale Southern Observatory

ABSTRACT: The present status of the Yale/San Juan Southern Proper Motion program (SPM) is described with emphasis on the PDS Input Catalog preparation. The SPM - a program which is specifically designed for galactic structure studies - will provide positions, absolute proper motions, magnitudes and colors for about one million stars in the southern sky. The input object selection/cross-identification procedure which has been developed at Yale is presented in detail. We encourage those with specialized catalogs of objects for which proper motion measures would be valuable to contact the authors for possible inclusion in the SPM input list.

1. INTRODUCTION

The Yale/San Juan Southern Proper Motion program (SPM) will provide absolute proper motions with respect to faint galaxies, positions, magnitudes and colors for approximately one million stars in the southern sky. Originally conceived as the southern extension of the Lick Northern Proper Motion (NPM) program, the purpose of the SPM has been broadened to include the creation of a secondary reference system of faint stars meeting growing demands from the astronomical community for such a system. In addition, the quickly expanding scope of various astrophysically-related catalogs gives an opportunity to provide absolute proper motions for certain, deliberately chosen classes of objects. The absolute proper motions coupled with other kinematical and astrophysical data will improve considerably our knowledge of the different components of the Galaxy.

The above considerations were incorporated in the PDS Input Catalog preparation strategy. What the SPM lacks in number of total objects, (relative to other sky-digitization programs), we hope to make up for in accuracy. Special attention will be given possible sources of systematic errors such as magnitude and color effects.

2. BASIC FEATURES OF THE SPM

A brief history of the SPM program has been described by Wesselink (1974) and van Altena et al. (1991). Here we simply summarize the basic information on the SPM plate

Workshop on Databases for Galactic Structure
A. G. Davis Philip, Bernard Hauck and Arthur
R. Upgren, eds. p. 153 - 159 © 1993 L. Davis Press

153

Imants Platais
Yale University Observatory
New Haven, Connecticut
06511

material. The sky coverage of the SPM consists of 623 fields (each 6.3 by 6.3 deg) south of -17°
on 5° centers. Besides the main area, there are an additional 94 SPM fields north of -17° which
eventually will be measured as well. A set of plates for each SPM field consists of two blue (B_{lim}
= 19 mag) and two yellow (B_{lim} = 18 mag) plates taken with a four-magnitude objective grating
and two hour and two minute offset exposures. Thus up to eight exposures per star per field
may be measured, times the number of measurable grating images (maximum of five) per
exposure. Approximately 200 fields currently have both the first and second epoch plates taken
at least 20 years apart.

The SPM plates are measured in object-by-object mode using Yale's 2020G PDS
microdensitometer upgraded with laser-interferometer encoders. The Yale Digital Image
Centering routine which fits a bivariate Gaussian to the image density distribution is used to
calculate image centers and other parameters.

3. RESEARCH PROGRAMS AND SOURCE CATALOGS

Both the NPM and SPM are ideally suited for Galaxy structure studies. Again, the main
research programs are extensively discussed by van Altena et al. (1990). These include various
aspects of galactic structure (galactic rotation; spatial and kinematical distribution of stars;
absolute proper motions of star clusters), fundamental astrometry, a faint secondary reference
frame (~ 300,000 stars within the range 16 - 18 visual magnitude, at a density of 15 - 20
stars/sq. degree) and, finally, the Hipparcos proper motion link to the extragalactic reference
frame.

Based on a three-component Galaxy model, a careful estimate was made of how many stars
one has to select in each magnitude interval to ensure statistically significant discrimination
between components (Mendez et al. 1992). Around the South Galactic Pole it is estimated that
a total of 31,000 anonymous stars between V = 7.5 and 18.5 mag should be measured across
40 SPM fields. This number of stars will allow one unambiguously to model the disk, thick disk,
and spheroid components from proper motions in the magnitude range covered by the SPM.

A subset of these anonymous, kinematic-studies stars are chosen to perform other tasks
also in the SPM reduction procedure. Relatively bright "bridge stars" are used to link the
systems of the long and short exposures, and well-distributed, faint anonymous stars are used
to perform the plate solutions, (see Platais et al. 1992 for more details). Added to this list of
anonymous stars should be all of the measurable galaxies, to provide the correction to absolute
proper motion. Other program stars are selected from a variety of source catalogs. The IRS,
ACRS and PPM catalogs will provide astrometric reference stars. Also, an attempt is made to
include all of the SIMBAD database objects above our plate limit as this is the most
comprehensive compilation catalog available. Finally, a number of special catalogs having
auxiliary data on certain types of stars (i.e. photometry, metallicity, radial velocities),
particularly around the SGP, are included in our input list in whole or in part. These will also
provide pure samples to determine the dispersions of proper motions for different types of
stars. The apparent magnitude distribution of all available objects in the SGP region as well as
the distribution of the SPM program objects are shown in Fig. 1.

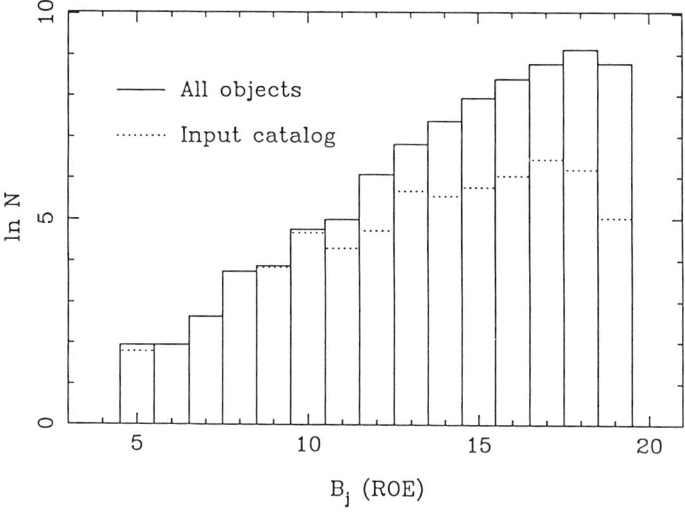

Fig. 1. A logarithmic histogram showing the magnitude distribution of all objects in the SPM field 528 (solid line) and the PDS Input Catalog (dotted line).

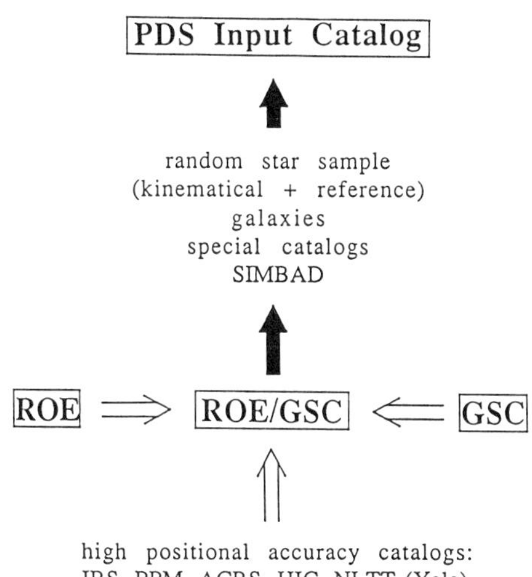

Fig. 2. Schematic flow-diagram of the PDS Input Catalog creation. Hollow arrows show the input data stream, bold - the selection process using the ROE/GSC master catalog.

4. PDS INPUT CATALOG

To maintain the highest attainable accuracy and keep the measuring time reasonable the total number of stars per SPM field in the PDS Input Catalog must be limited to a few thousand. Fig. 2 schematically shows the PDS Input Catalog creation guidelines. An input list of the SPM program stars is prepared with the aid of two large catalogs: the COSMOS/UKST Object Catalog of the Southern Sky (which we shall refer to as the ROE catalog), a deep-sky survey ($B_j < 21$ mag) which includes positions, magnitudes and object classification; and the HST Guide Star Catalog (GSC catalog), a less-deep survey ($B_{j\,lim} = 15.6$ mag at the South Galactic Pole) with similar information.

The ROE and GSC catalogs are combined to form a master list of stars and galaxies, virtually complete down to $B_j = 19.0$ mag. The ROE positions are tied into the GSC system using stars common to both catalogs. For many bright stars, more accurate positions (and proper motions) are adopted from catalogs such as the IRS, ACRS, PPM, etc. This ROE/GSC master catalog, it should be noted, is intended to support certain requirements in the pre-scanning stages of the SPM program and is not intended as a catalog of astrometric quality. It is invaluable in serving numerous functions in the preparation of the PDS Input Catalog: i) it is used as a base from which to select randomly the necessary anonymous stars as described above, ii) it serves as a master index for matching/cross-identification with special catalogs or objects of interest extracted from SIMBAD, iii) it provides a list of galaxies, necessary for the link to absolute proper motions, iv) it allows us to predict potential image crowding/confusion problems due to the multiple exposures and grating images on the SPM plates. With regard to this last point, a model of all images of a potential target star and its surrounding field is made to predict possible blends. Randomly selected (anonymous) stars are actually selected from those stars which do not have potential blend problems. It might also be noted that the ROE/GSC master catalog is useful for identifying objects of interest for which only a finding chart is available.

For all program stars, the identification procedure is by position and magnitude coincidence and is almost fully automated. Tolerances for the position and magnitude matching depend on the source catalog being matched to the ROE/GSC. In the case of inhomogeneous positional catalogs like SIMBAD, we divide the sample into high and low positional accuracy parts and perform the matching separately. Each identification carries just two numbers - ROE number and the original catalog number. All cases of non-matches and multiple matches are recorded in special diagnostic files along with the statistical information on the identification process.

Possibly, the most important feature (for our purposes) of the ROE catalog is its object classification of stars and galaxies. After a visual check on several SPM plates we find that nearly all visible and measurable galaxies are also in the GSC catalog despite the fact that the nominal cutoff for the GSC is at least two magnitudes above the limit of our plates! The galaxies are not necessarily GSC-classified as nonstellar objects and their estimated magnitudes are two to four magnitudes brighter than the ROE estimates. ROE versus GSC Bj-magnitudes for a sample of confirmed galaxies in a single SPM field are presented in Fig. 3. In general, we find the ROE magnitudes better agree with our magnitude estimates from the SPM blue plates. Before scanning a set of SPM plates, we check visually all "possible galaxies" selected from the

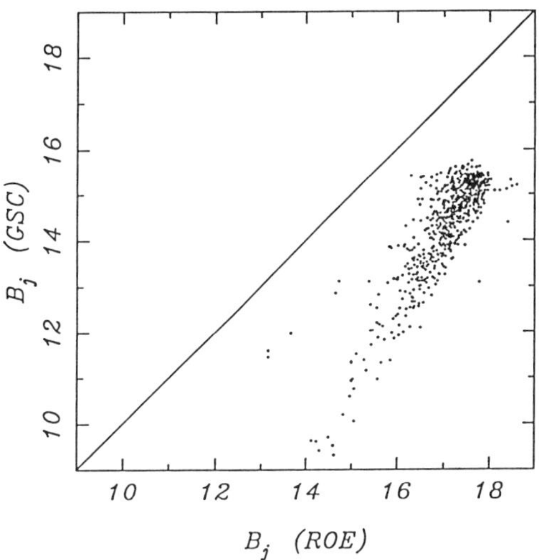

Fig. 3. ROE vs. GSC B$_j$-magnitudes for a visually confirmed sample of galaxies in SPM field 528 near the South Galactic Pole.

ROE/GSC master catalog directly on the PDS in order to keep only confirmed, measurable galaxies.

Once the SPM program stars have been selected and added to the input list, the PDS Input Catalog is created. Three calibrations are required to create the PDS Input Catalog, namely i) a transformation of the ROE/GSC coordinates into the PDS x,y-coordinates; ii) a transformation to produce both long and short exposure positions; iii) a star-diameter calibration as a function of ROE Bj-magnitude, to optimize the individual scan rasters. As a result of these calibrations the scan raster location and size are specified for every object in the input list and the scan can be commenced. In all cases when proper motions are available, the Input Catalog positions are upgraded to the plate epoch.

5. CURRENT STATUS

Provisional standard error of the proper motions derived from several SPM fields around the South Galactic Pole is four to five milliarcseconds per year (mas/yr) for stars < 16th visual magnitude. The uncertainty in the absolute proper-motion zero-point is 0.5 - 1.0 mas/yr depending on the number of measurable galaxies in the field. The internal precision of the equatorial coordinates is 0.10 arcsec (s.e.) per coordinate and per plate, at the plate epoch. The photometric accuracy of star magnitudes is ±0.15 mag per plate. Currently, we are in the process of measuring about 40 SPM fields around the SGP where the set of both first and second epoch plates is complete. After careful analysis of possible systematical errors by a variety of methods, the results of the extended SGP region will be made available to the astronomical community.

6. ACKNOWLEDGMENTS

The authors are pleased to thank Dr. Harvey MacGillivray for furnishing the relevant portion of the COSMOS/UKST Object Catalog of the Southern Sky. We are indebted to the members of the Astronomical Society of New Haven - C. Predom, R. DeMartino, M. Dzubaty, D. Kocyla and E. Wetherbee for their invaluable contribution to this project, improving the celestial coordinates of variable and NLTT stars. We also wish to thank R. Mendez for numerous helpful discussions and for providing and running the Galaxy model code used to establish criteria concerning the number of program stars. The SPM project is supported by grants from the National Science Foundation.

REFERENCES

Mendez, R. A., Platais, I., Girard, T. M. and van Altena, W. F. 1992 in VII IAU Latin American Regional Meeting, Revista Mexicana Astron. y Astrofis, in preparation

Platais, I., Girard, T. M., Mendez, R. A., van Altena, W. F., Lopez, C.E., MacGillivray, H. T., Yentis, D. J., Lattanzi, M.G. and Lasker, B. M. 1992 in Astronomy from Large Databases II, A. Heck and F. Murtagh, eds., ESO, Hagenau, p. 455

van Altena, W. F., Girard, T. and Lopez, C. E. 1991 in IAU Colloquium 100, Fundamentals of Astrometry, H. Eichhorn, C. A. Murray and A. R. Upgren, eds., Astrophys. Sp. Sc. 177, 59

van Altena, W. F., Girard, T., Lopez, C. E., Lopez, J. A. and Molina, E. 1990 in IAU Symposium 141, Inertial Coordinate System on the Sky, J. H. Lieske and V. K. Abalakin, eds., Kluwer Academic Pub., Dordrecht, p. 419

Wesselink, A. J. 1974 in IAU Symposium 61, New Problems in Astrometry, W. Gliese, C. A. Murray and R. H. Tucker, eds., Reidel, Dordrecht, p. 201

Yentis, D. J., Cruddace, R. G., Gursky, H., Stuart, B. V., Wallin, J. F., MacGillivray, H.T., Collins, C. A. 1992 in Digitised Optical Sky Surveys, H. T. MacGillivray and E. B. Thompson, eds., Kluwer Academic Pub., Dordrecht, p. 67

DISCUSSION

LU: The 31,000 stars centered at the SGP; do you have the number of stars per square degree? The star count in our class project using the UK Schmidt film to 21 mag is about 722 stars per square degree. The F and G star count in the SGP survey is about 78 stars per square degree between 8 to 16.5 mag.

PLATAIS: The total number of stars, 31,000, refers to approximately 40 SPM fields. That of course is just a small portion of all available stars. Due to limitations we define how many stars we want/can measure per plate and per a given magnitude interval.

LASKER: Your efforts to produce good positions for, objects (astrophysically interesting) with poor positions (GCVS, NLTT, etc.) are very important. How will this material be distributed?

PLATAIS: First, we plan to complete the regions around the SGP and publish the results as soon as possible. The catalog, after all necessary checking, will be available at major data centers. A few lists of improved positions for the GCVS stars have already been published.

COHEN: Is anyone making systematic efforts to supply spectral types for stars not contained in specialty catalogs but that you have in your Input catalogues? For example, SIMBAD may show that an object's spectral type goes back to the original coarse HD type only, no modern MK classification may exist.

PLATAIS: We would love to know of such an effort. As to our knowledge no systematic surveys (except the Michigan Survey) are under way.

CREZE: The proper motions you will get in the Magellanic Clouds will be ideal to check on the differential errors, such as the magnitude equation. Do you have plans to use this information?

PLATAIS: Yes, the Magellanic Clouds, as well as star clusters, are an ideal means to check the presence of such systematic errors. Regarding the Magellanic Clouds, a severe image crowding may cause a problem.

LASKER: With regard to your comment of the ROE and GSC galaxy data, the systematics is very much what we expected, but the small envelope is noted with some satisfaction.

PLATAIS: Both ROE and GSC are complementing each other. Since different approaches were used to determine magnitudes for galaxies, systematic differences of that kind should be present. Surprisingly, these differences are actually helping us to make a clear sample of galaxies for the SPM input list.

THE USNO PRECISION MEASURING MICRODENSITOMETER PROJECT TO MEASURE THE TWO EPOCH PLATE SETS OF THE PALOMAR OBSERVATORY SKY SURVEY

Jeffrey R. Pier and David G. Monet

U. S. Naval Observatory, Flagstaff Station

1. INTRODUCTION

The second Palomar Observatory sky survey (hereinafter POSS-II) (Reid et al. 1991) will provide the opportunity to study the entire northern sky at a higher resolution and to a deeper magnitude limit than the original POSS-I (the National Geographic Society-Palomar Observatory Sky Survey). Additionally, of course, comparisons of the two surveys, taken approximately forty years apart, will generate catalogs of enormous benefit to the study of galactic structure.

The California Institute of Technology and the U. S. Naval Observatory (USNO) have entered into an agreement for the USNO Flagstaff Station (NOFS) to digitize and reduce POSS-I and POSS-II plates for northern fields down to the equator (Monet 1993). To attempt to tie the astrometric reference frame of bright stars into the deep POSS plates, an additional, a short exposure (3 minutes), unbaked IIIa-J plate will be taken with the Oschin Schmidt telescope within a month of the respective POSS-II J plate.

The U. S. Naval Observatory's (USNO) Precision Measuring Microdensitometer (PMM) was designed to digitize photographic plate material at the highest speed possible while achieving an accuracy no worse than the systematic errors involved in the calibration of plates taken by the Oschin Schmidt telescope. This paper will discuss the PMM system and its current status.

2. THE PMM HARDWARE

The PMM was built by the Anorad Corporation of Hauppauge, NY, a manufacturer of large industrial milling machines. A carriage holds a stage which, in turn, holds the plates. The stage is large enough to carry four 14-inch POSS plates to allow for long periods of uninterrupted measuring. The carriage and stage ride on air bearings over a lapped granite slab. The slab is approximately 3 by 4 by 0.4 meters in size. The entire system weighs about 10 metric tons. The PMM is seen in a photo which appears as Fig. 1.

Carriage motion, driven by a torque motor and screw, is in the X direction. The carriage Y position is defined by air bearings which ride up against a granite beam attached to the base slab. Inside the carriage, a stage, driven by a brushless DC motor, carries the plates in the Y direction. The stage's X position is defined by air bearings held against a granite beam on the

Workshop on Databases for Galactic Structure
A. G. Davis Philip, Bernard Hauck and Arthur
R. Upgren, eds. p. 161 - 166 © 1993 L. Davis Press

161

Jeffrey R. Pier
U. S. Naval Observatory
Flagstaff Station
Flagstaff, Arizona
86002

carriage. The X and Y positions are laser encoded to 0.1 μ. A factory provided 80286 computer senses the laser-encoded positions and commands machine motions. Carriage and stage motion is variable and can reach up to 40 mm/sec. The cameras are mounted on the upper Z stage. The Z position is encoded to 1.0 μ and the distance between the upper plate surface and the Z stage is monitored with a laser distance sensor. Fixed illuminators are located below each camera.

Fig. 1. The PMM

Since there are a large number of plates (approximately 5000) to be scanned in a finite amount of time, the PMM design was driven towards the use of array detectors. Videk MegaPLUS CCD cameras were chosen to fit the bill. They are large format cameras (1320 x 1035 pixels), are blemish-free, have square contiguous 6.8 μ pixels, and are commercially available. With 2:1 reimaging optics, the projected pixel size is 13.6 μ (0.9 arcseconds on a POSS plate). Each camera has an internal mechanical shutter and an on-board 8-bit analog-to-digital converter.

3. DATA FLOW AND PROCESSING

Each "footprint" is comprised of about 1.3 MPixels. There are roughly 700 footprints per plate and approximately 5000 plates. Thus, once the 8-bit pixel data are converted into 32-bit floating point numbers, the flattened data frames will approach 20 terabytes of data.

The basic processing scheme is to
    1. STEP (position the plate);
    2. STOP;
    3. GRAB (expose, A-to-D conversion);
    4. PROCESS.
    repeat until done.

The digitized data are read out of the camera at 10 MPix/sec (about 1/7 second per frame). A fiber optics link connects the PMM cleanroom to the computer room where the data are read into a Silicon Graphics 4D/240S computer via a high speed VME bus. At present, the camera is read four times at each position on the plate and the four frames are averaged in Silicon Graphics memory.

The average frame is flattened using predetermined pixel-by-pixel response functions. Then a median filter algorithm is used to determine local sky and the noise in the sky. A "blob" finding algorithm is used to find all pixels above sky at some significance level, and traditional moment analysis is conducted on each blob to provide initial first guesses at position, size and shape parameters. The first guesses are then fed into more sophisticated astrometric, photometric and image classification algorithms.

The chosen astrometric algorithm is a non-linear least squares fit of a Gaussian function to the image marginal distributions. Photometric and image classification algorithms are under active investigation and development. At the present, without sophisticated photometry and image classification, data acquisition and analysis for each footprint takes roughly 6 seconds. Since there are roughly 700 footprints per plate, one plate can be processed in a little over an hour.

There are several computers involved in the system. User interface is via an X-Windows environment on a graphics workstation. PMM motion control and position determination is performed by the factory provided "Anomatic" 80286. Data acquisition and processing is performed by a pair of Silicon Graphics 4D/240S (one for each of the two cameras). Finally, a DEC MicroVAX-II is employed as the system synchronizer. The MicroVAX passes commands and status information to the appropriate machines and then monitors all activity to ensure that all activities are synchronized.

4. THE PRODUCT

The PMM program will result in the production of two catalogs. An internal catalog will be generated on a plate-by-plate basis. It will represent the contents of each plate, calibrated only to remove the instrumental signature of the PMM. This catalog will contain position, brightness and classification parameters for each detection.

Using available calibration data, the external catalog will contain coordinates (R.A. and Dec), magnitudes, colors, proper motions, plus additional image classification parameters for non-stellar objects. It is expected that the external catalog will contain approximately 10^9 entries, but this is somewhat difficult to estimate at this time. The success rate that can be achieved in correlating objects among plates and between epochs is not yet known. Some fields, at low galactic latitude for example, will undoubtedly prove to be difficult to handle. Since astrometry, including proper motions, is of particular importance to NOFS, it is anticipated that a major effort will be expended in list correlation. NOFS hopes to be able to make the external catalog widely available to the astronomical community, but, presently, funding has not been made available to accomplish this.

5. CURRENT STATUS (May 1993)

Basically, all of the hardware seems to work. The integration of the PMM, cameras, computers, data links and interfaces was surprisingly straight-forward. A prototype lightweight yet rigid plate holder, made of hexcore aluminum, is being evaluated. Other materials are being considered. Numerous checks and tests are being conducted to evaluate machine performance. Tests of positioning repeatability (obtain footprint, move to a new position in random direction, move back to original position, obtain footprint, compare star positions, repeat) and positioning determination of stars appearing on overlapping footprints indicate that the machine metrology is good to about 0.3 μ (0.02 pixels). Optical terms/distortion in the camera and camera optics have been reliably mapped at the 0.05 μ level using cubic polynomials. Changes of plate scale as a function of camera-to-plate (Z) distance are now well determined. Not surprisingly, the PMM metrology is quite sensitive to room temperature fluctuations and recently significant improvements in the cleanroom temperature stability have been obtained.

The data acquisition software is in good shape. A sequencer language has been developed for the MicroVAX which serves as an interpreter for commands, macros and sequences of commands which are sent to the respective computer to move the machine, read machine coordinates, take a frame, process the frame while moving machine to next position, etc. As far as the real-time data analysis software goes the image detection and astrometry codes are in good shape (NOFS has a lot of experience in performing astrometric measurements on digitized images). We are still evaluating algorithms to handle overlapping/extended images, object classification and photometry.

Follow-on analysis software development is in very preliminary stages. Major remaining projects include such things as list correlation of large numbers of objects, defining catalog parameters, defining the database and writing the database management system.

6. TIMELINE

The remainder of 1993 will be spent measuring test plates, continuing with the analysis software development and beginning the catalog preparation software development. We hope to "measure" our first plate by the end of the summer. A second contract Full Time Equivalent scientist will be arriving this summer and we will be upgrading the Silicon Graphics computers from 4D/240s to 4D/440s which will mean about a 60% increase in CPU speed.

In 1994 we expect to begin an initial trial production measurement. We will also, of course, continue with the development of merging software and will import the databases which will be used for the astrometric and photometric calibration of the plate measures. We expect to produce a merged, calibrated catalog of a small area in 1994.

After we analyze these data we will need to make some hard and lasting decisions. To produce a uniform catalog of the whole sky we want to conduct all measurements in the same fashion, i.e. no significant changes in hardware, scanning techniques, data acquisition and/or real-time data analysis can be tolerated during the multi-year project. We anticipate that we will need to restart the measures at least once, and we probably would not be surprised if two or three restarts were required before things really get rolling.

ACKNOWLEDGEMENTS

The PMM project involves the support of the entire NOFS staff, but particular thanks are owed to Harold Ables, Alan Bird, Hugh Harris, Arne Henden and Sandy Leggett. Thanks are also due to the leadership of the USNO in Washington: the USNO Scientific Director Gart Westerhout and the four recent USNO Superintendents: Captains Charles Roberts, Richard Anawalt, James Hagen and Winfield Donat. The measurement of the POSS-I and POSS-II plates is made possible through the cooperative support of several Caltech scientists; we especially thank Bob Brucato, Judy Cohen, Jeremy Mould, Gerry Neugebauer and Maarten Schmidt.

REFERENCES

Monet, D. G. 1993 in Sky Surveys: Protostars to Protogalaxies, B. T. Soifer, ed., ASP Conference Series, Volume 43, p. 139

Reid, I. N., Brewer, C., Brucato, R. J., McKinley, W. R., Maury, A., Mendenhall, D., Mould, J. R., Mueller, J., Neugebauer, G., Phinney, J., Sargent, W. L. W., Schombert, J. and Thicksten, R. 1991 PASP 103, 661

DISCUSSION

LASKER: What are your intentions with respect to the image data (pixels)?

PIER: The pixels will be saved locally in case it becomes necessary to reanalyze any of the frames, but, unfortunately, we don't have the manpower or resources to make the pixels widely available to the astronomical community. We're counting on STScI to provide that service.

VAN ALTENA: Is the USNOFS Quick J survey filtered or unfiltered?

PIER: It is filtered the same as the POSS-II J plate (GG 385) but the plates are not baked.

CASERTANO: Since you expect the repeatability of your measures to be better than 0.3 μ, can you comment on the total error figure of 1.5 μ (0".1) you quoted?

PIER: 1.5 μ was the design goal for astrometry. Current metrology is good to 0.3 μ on a single plate in a single orientation. We don't yet know what the final measuring error will be, although we are hopeful that we can do better than the design goal.

APPLICATION OF MAXIMUM LIKELIHOOD TO THE ANALYSIS OF LARGE STELLAR CATALOGS

Stefano Casertano

University of Illinois

Kavan U. Ratnatunga

Johns Hopkins University

ABSTRACT: We discuss a numerical algorithm based on maximum likelihood for estimating parameters for models of the Galaxy. We use simultaneously all the information available in a catalog of stellar data in a global optimization, to derive unbiased estimates of intrinsic stellar properties, such as absolute magnitude and kinematics. The likelihood function is defined in the observed domain using quantities such as photoelectric photometry, line-of-sight velocity, proper motion, trigonometric parallax, and metallicity. Individual stars included in the statistical analysis can have different amounts of information available. This method includes an explicit treatment of observational errors, can identify outliers objectively and allows use of stellar data with relatively large errors. It can self-consistently detect and correct for systematic deviations in the observations, such as zero point residuals or underestimated errors.

1. INTRODUCTION

The study of stellar statistics has a long and distinguished tradition. Since the days of Hipparchus, a major task of astronomers has been to compile star catalogs as large and as accurate as possible. Over the last decade many large databases have been compiled from both space and ground based observations. It can be expected to improve further in size, precision and uniformity given by automated survey telescopes.

The traditional approach of statistical astronomy given in Trumpler and Weaver (1953) has seen only few advances over the next four decades of astronomical research. Among them, particularly for the problem of calibrating the luminosity of stars with measured parallaxes, are Jung (1971), who developed a maximum-likelihood scheme to deal with complete, magnitude-limited samples, and Hansen (1973), who developed a method for non-linear optimization of the spectroscopic calibration. Turon Lacarrieu and Crézé (1978) first suggested the use of magnitude constraints to remove the divergence found by Lutz and Kelker (1973) in parallax-based distance estimators. Smith (1988) wrote a series of papers on the calibration problem, specifically discussing the advantages of maximum likelihood in this context, and gave an application to synthetic data.

As computation time was the major constraint, all these methods were based on analytical equations derived using clever approximations, and assumptions about the distribution function

Workshop on Databases for Galactic Structure
A. G. Davis Philip, Bernard Hauck and Arthur
R. Upgren, eds. p. 167 - 171 © 1993 L. Davis Press

Stefano Casertano
Physics and Astronomy Dept.
505 Bloomberg Bldg
The Johns Hopkins University
Baltimore, Maryland
21218

and observational error. For example, the standard χ^2 test assumes that error distributions are Gaussian. In this case the likelihood function, defined as the sum of the natural logarithm of the probabilities for individual stars, equals $-1/2\chi^2$ + constant. The χ^2 minimization is therefore equivalent to a special case of maximum likelihood. The principle of maximum likelihood can however be used for arbitrary probability distributions, and can be implemented by evaluating numerically the individual probabilities for each element (star) from a computer model of the distribution function, which can then be as complex as necessary to model the observations.

In many cases, we are mostly interested in global properties of the stellar distribution. Therefore it is not necessary to estimate an individual distance to each star. We represent distance by a probability distribution over "true" distance, rather than a single mean value. At each "true" distance, a well defined galaxy model can always be transformed precisely to the observed plane. The likelihood function is then evaluated by numerically integrating the probability over "true" distance. We also need not classify each star into a particular stellar population. For example, a low velocity star, which most probably belongs to a population with low velocity dispersion, has a non-negligible probability of belonging to a component with high velocity dispersion. Attempts to bin the stars into stellar populations and analyzing them separately will bias kinematic parameter estimates. We evaluate the conditional probability of membership in each stellar component and sum over the mixture of stellar populations. This numerical approach is only feasible thanks to the increasing availability of fast computers.

2. GLOBAL MODELS

The most important advantage of using global models for studying the structure of the Galaxy is to avoid the transformation of observed quantities. Consider the estimation of the space velocity for a star. This involves at least three independent measurements: line-of-sight (radial) velocity; proper motion; and a distance estimator, which in turn may depend on one or more observables. Each measurement will contribute its own error, which is transformed in a complicated way into the derived quantity. Unless the errors are very small, their resulting distribution can be very complex, and is in general poorly behaved. Even worse, in some cases there may not be *any* well-defined transformation. For example, measurement errors can make the observed trigonometric parallax negative. Although a negative parallax cannot be transformed to a real distance, the probability of making this observation for a star at any "true" distance is well defined. Global modeling always brings the comparison between model and data into the observed domain, where the error distributions are expected to be simple.

Another useful advantage of global modeling is the flexibility it allows as to which data must be available. To continue the above example, if the line-of-sight velocity is not available for a star, the space velocity for that star cannot be computed, and thus any other data available for that star may be wasted. On the other hand, with global modeling the comparison can be made using only the available observables for each star, and thus all information in the database can be used.

Finally, the global approach can detect and compensate for selection effects in the sample. A numerical computation of the expected distributions of the observables can include the various selection criteria applicable to the catalog. A properly constructed model will show incompleteness with respect to any deficiency by direct comparison.

Among the main proponents of the global modeling approach to galactic structure problems have been Bahcall and Soneira (see references in Bahcall 1986). Following in their path, we produced the IASG model (Ratnatunga, Bahcall and Casertano 1989), which, thanks to improved software, has enabled us to study all-sky catalogs, including kinematics. The IASG model allows for selection limits to be imposed on any observed quantity and for tailoring a model sample of stars on an observed catalog on a star-by-star basis, thus all but eliminating the consequences of incompleteness of the sample. It has been used to study both the luminosity function and the kinematics of intrinsically bright stars, and to derive the kinematic properties of spheroid and old disk from high-proper motion stars.

The global approach described above is not without some disadvantages. The first is inherent in the process of comparison between model and data. Such comparison can only be done effectively in one observable; with more than one, the comparison becomes more difficult, since the data points cover only sparsely a multidimensional domain. The best that can usually be done is a study of the marginal distribution, or a limited comparison involving correlation coefficients or principal components. Any of these possibilities reduces the information actually used in the test.

Another difficulty is parameter optimization. The process of finding the best model must proceed by trial and error, with all the subjectivity that this entails. The adopted model is necessarily guided by parameters derived using the traditional approach, which studies the distribution of those quantities that, while not directly observed, have a more obvious physical meaning. The global model then essentially complements the classical analysis by checking self consistency of independently derived parameters.

3. MAXIMUM LIKELIHOOD

Both difficulties listed above can be overcome by using the method of maximum likelihood. With this approach, the global galaxy model is used to derive numerically the *probability* of a set of observables with associated errors. To correct for incompleteness in some observables, say for instance photometry and region of sky surveyed, we use *conditional probabilities* to estimate the probability of other observables, for example the kinematic properties, at the *observed* position, apparent magnitude and color for each star individually. This probability is interpreted as a measure of the *likelihood* of the model given that particular observation. The likelihood of the model given the sample is the product of individual likelihoods for each star in the catalog; in practice, we use its natural logarithm, the *likelihood function*. The best model is the one for which the likelihood function is largest. The matrix of second derivatives (Hessian) of the likelihood function at the maximum is the inverse of the covariance matrix of the model parameters.

There is essentially no limit to the complexity of the model, although the quality and quantity of data available will limit of course the ability to constrain many parameters. Several independent components can be defined; each will have its own density law, geometrical shape, luminosity function, color-magnitude relationship and cosmic scatter, mean motion (including overall motion, rotation, and expansion), and velocity dispersion.

The number of free parameters included to model the observed distribution is determined using the likelihood ratio test. The number of parameters is increased until there is no significant increase in the likelihood function. After estimating the Galaxy model parameters the observed distributions is compared with the global model. Outliers and deviations from the functional form of assumed distributions are detected at this stage. A non-Gaussian error distribution would lead to a poor match between model and data. Also, significant differences will arise in the presence of problems in the data, such as a zero-point error and/or underestimate of the observational errors.

With our method, the information content of each star is exploited to the fullest. No binning is used. Error estimates and correlations are derived for each parameter. Since an explicit cost function is computed for each model, the search for the *best* model is objective, not subjective. Good algorithms are available to locate the maximum; several have been tried, and different ones can be use for each problem to ensure optimum performance and to avoid false (local) maxima.

A practical application of the numerical maximum likelihood method to stellar statistics is due to Casertano et al. (1990), who determined the properties of the old disk population of the Galaxy from kinematics and photometry only, without any metallicity information and without any preconceived notion of its properties. We have subsequently used variants of this method, with different combinations of photometry and observables: parallaxes (Ratnatunga and Casertano 1991), metallicity and line-of-sight velocity (Ratnatunga and Yoss 1991), and proper motions and line-of-sight velocity (Ratnatunga and Upgren 1993). Application to study properties of nearby open clusters is described by Casertano et al. (1993) in this volume.

4. CONCLUSIONS

With personal workstations becoming as fast as main frame computers of a decade ago, statistical analysis of astronomical data need no longer be constrained by analytical approximations. Astrometric data has been costly to acquire in terms of human resources, instrumentation, and time and numerical algorithms are useful to extract unbiased parameters from them. We have developed one such maximum likelihood algorithm for the analysis of stellar data and we have tested it on existing ground-based catalogs in preparation for large uniform database from the Hipparcos satellite which will surely give a giant leap for galactic structure.

ACKNOWLEDGMENTS

Our work owes its initial impetus to John Bahcall, who initiated both authors to the mysteries of galactic modeling and instigated the use of maximum likelihood for such problems. We have benefitted during the years from discussions with many colleagues, and especially with Bruce Carney, Heinz Eichhorn, Ken Freeman, Ivan King, Mario Lattanzi, Art Upgren, Ken Yoss and Wayne Warren. We gratefully acknowledge NASA support through grant NAGW-2945.

REFERENCES

Bahcall, J. N. 1986 ARAA 24, 577
Casertano, S., Ratnatunga, K. U. and Bahcall, J. N. 1990 ApJ 357, 435
Casertano, S., Lattanzi, M. G. and Ratnatunga, K. U. 1993, in Workshop on Databases for Galactic Structure, A. G. D. Philip, B. Hauck and A. R. Upgren, eds., L. Davis Press, Schenectady, p. 229
Hansen, L. 1973 A&A 27, 355
Jung, J. 1971 A&A 11, 351
Lutz T. E. and Kelker, D. H. 1973 PASP 85, 573
Ratnatunga, K. U., Bahcall, J. N. and Casertano, S. 1989 ApJ 339, 106
Ratnatunga, K. U. and Casertano, S. 1991 AJ 101, 1075
Ratnatunga, K. U. and Upgren, A. R. 1993 ApJ, submitted
Ratnatunga, K. U. and Yoss, K. M. 1991 ApJ 377, 442
Smith, H. 1988 A&A 198, 365
Turon Lacarrieu C. and Crézé, M. 1977 A&A 56, 273
Trumpler, R. J. and Weaver, H. F. 1953 Statistical Astronomy, University of California Press, Berkeley

DEEP PROPER MOTION SURVEYS IN OPEN CLUSTERS. I. M 67

Mario G. Lattanzi[*]

Space Telescope Science Institute

ABSTRACT: Most of the present astrometric work on open clusters is severely limited in magnitude by relatively short exposure first epoch plates. This has hampered systematic investigations of clusters' lower main sequence and of the luminosity function of less massive (below 1 $M_\odot$) members. With the advent of fast scanning machines and improvements in reduction techniques it is now possible to make good astrometric use of early epoch material like the POSS-I sky survey. In this contribution I will discuss preliminary results for the old standard cluster M 67. The plate material used includes all relevant Schmidt plates available from the STScI digital archive, deep 1950 plates taken with the Palomar 200-in telescope and CCD frames as second epoch exposures.

This survey in M 67, which covers a region of about 20 arcmin on a side centered on the cluster, is primarily aimed at the confirmation, as cluster members, of a set of stars fainter and bluer than the main sequence turnoff (but above the white dwarf cooling sequence).

1. INTRODUCTION

Photometric surveys in galactic cluster fields have been successfully used since the early days of photographic astronomy. Their main objective is the study of the structure of color-color and color-magnitude (CM) diagrams through comparisons to synthetic diagrams based on the calculation of stellar evolutionary models. The availability of better plate emulsions and photoelectric photometers with large telescopes (from the early fifties) and, later, of CCD detectors has made the extension to faint magnitudes relatively easy. Deep photometric surveys have then been able to uncover, for example, the unevolved main sequence (MS) of distant clusters, as well as the low mass segment of the MS of closer clusters, which has dramatically improved cluster age and distance determinations.

However, no firm conclusion, based solely on photometric data, can be reached on, for example, the existence (and the physical properties) of a binary sequence which runs almost parallel to the usual (single star) MS, or the membership of astrophysically peculiar stars, or the number of white dwarfs (WD) really belonging to a cluster, until independent cluster membership is established for each candidate through the analysis of kinematic data. Both good radial velocities and proper motions should be used for best results but this opportunity has not been fully exploited so far because of the difficulty of gathering the necessary data (especially radial velocities). The situation gets worse when extending to fainter magnitudes as larger telescopes are needed.

[*]Affiliated with the Astrophysics Division, SSD, ESA; on leave from Torino Observatory.

Workshop on Databases for Galactic Structure
A. G. Davis Philip, Bernard Hauck and Arthur
R. Upgren, eds. p. 173 - 180 © 1993 L. Davis Press

Mario G. Lattanzi
Space Telescope Science Institute
3700 San Martin Drive
Baltimore, Maryland
21218

Proper motions alone are relatively cheap to obtain in large numbers thanks to modern plate measuring techniques (see, for example, Lasker 1993, this volume). The problem is that reliable proper motions for kinematic memberships in open clusters require cluster images taken decades apart. The best first epoch material is, in most cases, relatively short exposure plates which have limited virtually all proper motion studies to stars brighter than $V \approx 16$, i.e. to stars rather close to the MS turnoff. (A very restricted number of nearby clusters like the Hyades and the Pleiades are notable exceptions.)

M 67, which exhibits a rather complex CM diagram in connection with its advanced evolutionary state (age = 4.5 Gyr), has been the focus of three modern proper motion studies (Murray et al. 1965, Sanders 1977, and Girard et al. 1989). Although the quality of the proper motions has improved by a factor of five since the work of Murray and collaborators, the magnitude limit has remained practically unchanged at $V \approx 16$.

Because of the lack of faint proper motions some very interesting questions raised by the appearance of the deep CM diagram of Racine (1971) were left unanswered. What happens to the "binary star strip" which runs parallel to the MS between $V = 13.5$ and $V = 16$? Are any of the eight candidate WDs found in the visual region 18 - 22 bona fide members? Are any of the six stars below and to the left of the MS at $14 < V < 17$ (for which there are no proper motions in Murray's paper) astrometric members of M 67? Photometric errors are on the average better than 0.05 mag.

I will discuss some of those topics in the sections below (sections 4 and 5), but first a few notes on the new deep proper motions and proper motion-based cluster memberships (Sections 2 and 3).

2. PHOTOMETRY AND ASTROMETRY

CCD photometric data (B and V bands) used in this work have been kindly supplied by L. Drissen, R. Gilliland and M. Shara. Most of the frames used in the reductions were taken at KPNO with 2048 x 2048 CCDs. The images cover a field of 20' on a side centered on the cluster (the field center is the same as that of Murray et al. 1965 study).

The astrometric plate material used for this new proper motion study includes, as first epoch images, the two relevant O and E POSS-I Schmidt plates (plate epoch is 1951.91) and two Palomar 200-in telescope plates, a B 5 min exposure and an U 30 min exposure, both taken the night of Feb 17/18 1958. Plate N486 of the Quick-V collection (Lasker et al. 1990) (taken in 1983.12) and the CCD best V frame (1991.2) were the second epoch exposures. Plate scales are 67".2/mm, 11".1/mm, and 0".6/pixel for the Schmidt plates, the 200-in plates and the CCD frame respectively.

Digital copies (25 μm pixels) of the POSS-I and Quick-V plates were already available in the Guide Star Catalog digital archive. The two 200-in Palomar plates were both digitized with a pixel size of 15 μm using PDS1 at STScI. The suitability of the STScI machines for accurate proper motion work has already been established (Lattanzi et al. 1991). All astrometric measurements were performed by L. Drissen using the PSF centroiding technique of DAOPHOT (IRAF version). For the plates, it was decided to build an average PSF using isolated stars in the magnitude range $14 < V < 17$. Thus a little degradation of the measured positions was expected at the bright ($V = 13.5$) and faint ($V = 18$) ends of our sample. Typical centering errors are better than 0.1 pixel.

A master list was generated from the CCD frame. This list goes as faint as $V = 21$ but its completeness limit is around $V = 20$. A relatively small subset of objects with bad measurements (usually stars in more crowded areas of the cluster) was removed from the sample. The magnitude limit of the proper motion sample ($V = 18$) is set by the Quick-V plate which doesn't provide reliable positions for stars fainter than this limit.

To derive relative proper motions I initially transform the rectangular coordinates of all the images into a common reference frame as defined by one of the exposures (the CCD frame in this case). This requires a proper choice of the reference stars for the derivation of plate-to-plate transformations. From our previous proper motion solution (Shara et al. 1993) which did not include the 200-in plates, I have extracted a set of high probability cluster members to define the trial rest frame. The three Schmidt plates were reduced using independent first order polynomials for each coordinate axis and no color or magnitude terms. The absence of any higher order term is probably a consequence of the tiny fraction of a Schmidt plate involved in the reduction (square patches of only 1.7 cm on a side). In contrast, astrometry must be done very carefully with the 200-in plates because of various distorsions and the onset of coma. Following Chiu (1976), I have used an 11-constant model (one for each coordinate) to reduce the 200-in plates to the flat surface defined by the CCD frame. The model includes a full quadratic polynomial, selected third and fifth order radial distortion terms [$x (x^2 + y^2)$ and $x (x^2 + y^2)^2$ for the X-axis], a linear color term, and linear and quadratic magnitude (V) terms. It must be noted that the very existence of a tight CM Diagram for the cluster members makes the magnitude and color terms highly correlated and the removal of the magnitude equation more difficult. Scatter in the CM Diagram helps breaking that correlation which would disappear if a sample of field stars is chosen as reference set. The necessity of "flattening" the 200-in plates on to the CCD frame (using small sections of the old Schmidt plates as reference images would result in a further compression of the 200-in plate scale by a factor of 2.5) forced me to use cluster members as reference stars.

The procedure described is iterated to eliminate outliers among the reference stars. None of the solutions required more than two iterations to converge. The average post fit residuals of the 200-in plate solutions are better than 0.1 pixel$_{CCD}$ (i.e. 0"06). Plots of the residuals versus V and (B-V) do not show any significant residual systematics.

The derivation of the proper motions is now straightforward. A linear least squares fit (independently for each axis) is done to the transformed positions spaced in time according to the corresponding plate epochs. The slopes of the individual fits are the sought for proper motions. The (internal) proper motion errors are also easily computed in this way. Final proper motions are good to 0"002/yr (rms per coordinate) between $V = 14$ and $V = 16$. The error goes up to 0"003/yr for fainter magnitudes.

3. ASTROMETRIC MEMBERSHIP PROBABILITIES

The derivation of cluster memberships closely follows the procedure described in Murray et al. (1965). I first group the total proper motions into bins of one magnitude from V = 13.5 down to V = 18. For each magnitude bin the average proper motion error (rms) is calculated. In proper motion space, the cluster "center" is at (0,0). One can then draw concentric circles whose radii are fractions of σ_m, where σ is the proper motion error corresponding to the magnitude bin m as defined above. I have defined four membership classes (one, two, three and zero) with one representing the most reliable members and zero definitive non-members. These membership classes correspond to radii of 2, 2 x $\sqrt{2}$, and 2 x $\sqrt{3}$ σ_m. This way the circle and the two annuli have equal area in the proper motion plane, so that the contribution of field stars, if any, to the total number in each of these three areas should be roughly the same. Class one members are those falling inside the first circle, class two members those in the annulus between the first and the second radius, and so on.

To put those membership classes into an absolute scale of membership probabilities I compared my classifications with that of Girard et al. (1989), which rely on proper motions a factor of three better than those discussed here. Of all the stars with membership probability > 90% in Girard's list 33 are in common with ours. At the other end of the probability distribution, I have 20 stars in common with Girard's low probability members (probability < 10%). Twenty seven of the 33 high probable members were in my class one circle, and 17 of the 20 non members are outside the 2 x $\sqrt{3}$ σ_m annulus. Roughly, our class one cluster membership corresponds to a membership probability of 82%.

4. THE CM DIAGRAM

Fig. 1 shows the CM Diagram of M 67 based on photometric data and astrometric memberships as discussed in the previous two sections. Only class one (filled circles) and class two (open triangles) have contributed to Fig. 1. It is probably the most accurate CM Diagram ever for this fundamental cluster and extends more than two magnitudes fainter than that given in Racine (1971).

This CM Diagram shows clear evidence for a large population of equal-mass unresolved binaries which runs in a strip parallel to the MS above V $\approx$ 15 and below V $\approx$ 16. Similar sequences are known to occur in other well studied clusters like the Pleiades and Praesepe. However, the M 67 MS appears very tight for 14.7 < V < 15.5, arguing for a "binary star gap" in this magnitude range, a finding that certainly deserves further investigation.

Star No. 600 in our master list (III-79 for Eggen and Sandage), a W UMa system discovered by Gilliland and collaborators (1991) is confirmed as a member of the cluster as determined from its proper motion. This brings the number of W UMa contact binaries in M 67 up to three. The frequency of such stars in M 67 is now rather close to that of NGC 188. Thus, NGC 188, with its seven W UMa stars, is no longer special. The high frequency of W Uma systems in the old clusters M 67 and NGC 188 relative to the field supports the idea that the number of close binaries in clusters relative to single members half the mass of the binaries can grow as the single stars are less bound to the cluster and more likely to be lost to the field (Gilliland et al, 1991).

If we are to take astrometric memberships seriously then, for the first time, we have to explain what are those four class one members with 15.5 < V < 18 and 0.45 < (B-V) < 0.95,

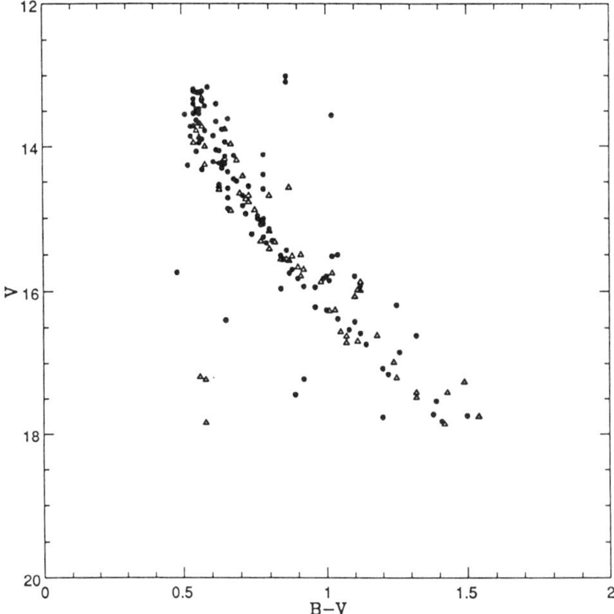

Fig. 1. CM Diagram of M 67 based on newly derived astrometric membership probabilities

which were named *blue interlopers* (BI) in Shara et al. (1993). That is a forbidden region of the CM Diagram as predicted by conventional stellar evolution theory. We discuss this in the next section.

5. ARE BLUE INTERLOPERS REAL?

Kaluzny (1990), after subtracting the field contribution from his CCD photometry of NGC 188, noticed that a significant number of "blue stars located up to two magnitudes below blue struggler region" remains in his cleaned CM Diagram of the cluster. A similar photometric "excess" was found in M 67 (Shara 1993), which has been somewhat validated by this investigation.

The simplest explanation for the BIs in Fig. 1 is that the proper motions are in error and those are merely field dwarfs. But it is not easy to discard class one members. As discussed in Section 3, this corresponds to a membership probability of more than 80%. Thus, probably only one of the four class one BIs is a non-member. On the other hand, class one below V = 17 is probably less accurate than class one at brighter magnitudes. All five class one BIs reclassified here were in the earlier sample of 12 class one BIs described in Shara et al. (1993) Then, as expected, more accurate proper motions have produced only rejections and no new detections. However, the number of primary BIs has dropped in a way which is consistent with an increased proper motion precision of a factor of three (in the variance). This would indicate that field contamination (the separation of the cluster from the field in proper motion space is indeed very difficult for M 67) has generated the observed BIs. We will not know what the right explanation is without further improving on the proper motions. Better yet, Echelle spectroscopy should provide the data (accurate radial velocities, spectral classification and metallicity analysis) for a definitive membership classification of the 8 candidates BI.

Efforts are underway to gather all the necessary spectroscopic data.

6. SUMMARY AND CONCLUSIONS

We have presented preliminary results on the deepest proper motion survey to date of the old open cluster M 67. The CM diagram shows clear evidence for a large population of equal mass binaries above $V \approx 15$ and below $V \approx 16$. The MS is very tight for $15 < V < 16$ arguing for a binary "gap" in this magnitude range. A third W UMa contact binary (III-79) candidate in M 67, discovered by Gilliland and collaborators (1991), is a confirmed astrometric member of the cluster. Finally, it remains to be seen if the seven blue interlopers left after applying the proper motion membership sieve are real members of the cluster or a statitistical excess generated by the field. Spectroscopic observations are underway to address this crucial issue.

ACKNOWLDEGEMENTS

I am indebted to L. Drissen, R. Gilliland, and M. Shara for letting me use some of their photometry in advance of publication and for their help in the reduction of CCD and plate data. I also wish to thank M. Potter for the plate scanning and G. Massone for his assistance with the proper motion code.

REFERENCES

Chiu, L. -T. G. 1976 PASP 88, 803
Eggen, O. J. and Sandage, A. R. 1964 ApJ 140, 130
Gilliland, R., Brown, T. M., Duncan, D. K., Suntzeff, N. B., Lockwood, G. W., Thompson, D. T., Schild, R. E., Jeffrey, W. A. and Penprase, B. E. 1991 AJ 101, 541
Girard, T. M., Grundy, W. M., Lopez, C. E. and van Altena, W. F. 1989 AJ 98, 227
Kaluzny, J. 1990 Acta Astronomica 40, 61
Lasker, B. M. 1993 in Workshop on Databases for Galactic Structure, A. G. D. Philip, B. Hauck and A. R. Upgren, eds, L. Davis Press, Schenectady, p. 77
Lattanzi, M. G., Massone, G. and Munari, U. 1991 AJ 102, 177
Murray, C. A., Corben, P. M. and Allchorn, M. R. 1965 R.O.B., No.91
Racine, R. 1971 ApJ 168, 393
Sanders, W. L. 1977 A&AS 27, 89
Shara, M. 1993, private communication
Shara, M., Lattanzi, M. G., Gilliland, R., Drissen, L. and Morse, J 1993 STScI Mini-Workshop on Blue Stragglers, R. Saffer, ed., ASP Conf. Series, in press.

DISCUSSION

ROESER: Just a comment: I think your conclusion about the five stars is quite all right. M 67 is a rather difficult case to disentangle the cluster members from possible background stars.

LATTANZI: That is certainly true! The compelling reason for improving the proper motions was their increasing ability to isolate the cluster from the field stars.

ROMAN: How bright are the stars below the main sequence? Spectra should show whether they are members - both from the radial velocities and from the appearance of the spectra.

LATTANZI: The location of the candidate blue interlopers in the CM Diagram is roughly $15 < V < 18$ and $0.45 < (B-V) < 0.95$. As I mentioned at the very beginning of my talk, we have an on-going spectroscopic campaign (led by M. Shara in collaboration with D. Latham, J. Morse, and R. Gilliland) which is intended to provide spectra for classification, metallicity estimates, and radial velocities. At the moment we have MMT Echelle high resolution spectra for three of the twenty-three candidates included in the original list. Only one of them is among the four high probable blue interloper candidates left after this new study. MMT spectra should provide us with radial velocities good to one km/sec. Also, radial velocity observations will be repeated as often as possible to investigate the presence of binaries in our sample.

MERMILLIOD: There is one possibility that stars located to the left of the main sequence are members: if the K dwarf has a white dwarf companion. Therefore you need several radial velocities to check the membership because they are binaries. Two examples are known in the Hyades cluster.

LATTANZI: This is certainly an attractive possibility for blue interlopers in M 67. Unless the appearance of the spectra (as for example a weak-lined spectrum indicating a metal content inconsistent with that of the cluster) suggests to drop some of the candidates, we will indeed wait until we have at least more than one radial velocity per candidate before estimating radial velocity memberships.

GRIFFIN: Could you please comment on the usefulness to this type of study of older plate files, though not as deep? I'm referring in particular to the Sonneberg plate files, taken at Sonneberg Observatory in former East Germany. People here may known that Sonneberg is threatened with closure in about 18 months' time, and the more the potential and actual value of the Sonneberg plates are publicized now, the greater may be the chance to enable the Sonneberg team to continue its good tradition.

LATTANZI: As I am not familiar with the Sonneberg plates I cannot directly answer this question. However, as my experience with other plate archives (like the Carte-du-Ciel archives of Cape Town, Cordoba, Perth and The Vatican) has shown, if the material has been properly preserved it is probably very useful in investigations related to open clusters and its value increases, of course, with time. If, for example, the plates cover a few square degrees then they can help the identification of cluster members evaporating from the cluster. This is crucial in cluster luminosity function studies as there are claims that, for example, in small and medium size young open clusters less massive stars are missing, which would lead to the conclusion that star formation in those sites discriminates toward small masses. Also, the Sonneberg plates could substantially help in improving the optical reference frame (positions and proper motions), which lacks stars fainter than $V = 10$. We should not forget than neither HIPPARCOS nor TYCHO will solve this problem, which I believe is one of the outstanding open issues in catalog astrometry.

LU: Do you know what membership is in for those stars below the main sequence of your color-magnitude diagram? Membership may be uncertain if the membership sigma is 3 σ or 4 σ.

LATTANZI: Astrometric membership probabilities, as computed in this study, of all the stars shown in the CM Diagram are higher than 0.80.

GUO: In your transformation model, you adopted linear and quadratic magnitude terms. Do you think these two terms are enough to take care of the magnitude equation? Do you have a plot of what transformation residuals verses magnitude?

LATTANZI: Yes, I believe that the magnitude equation has been removed to a sufficient accuracy with the inclusion in the least squares adjustment of linear and quadratic terms in V. The plots of the post fit residuals and of the proper motion errors versus V shown earlier clearly indicate that anything left of the magnitude equation is well below the random part of the proper motions mean error.

UNDERSTANDING RADIO SURVEYS OF GALACTIC H I

Felix J. Lockman

National Radio Astronomy Observatory
Greenbamk, West Virginia

ABSTRACT: The numerous galactic H I surveys that have been made over the past decades have usually been released to the community in a convenient digital form. The data, however, are rarely sufficient in themselves, but require some amount of interpretation and frequently a large amount of caution before they can be used successfully. This article discusses some reasons for this state of affairs and provides an introduction to those who might want to use 21 cm H I data bases in their research.

1. INTRODUCTION

This article is for those who wish to have some knowledge of the properties of atomic hydrogen in our Galaxy. Table 1 gives parameters of some recent surveys of the 21 cm emission line (this complements the list in Burton 1988). Most of the surveys have a particular distinguishing feature such as high angular resolution (the Arecibo survey of Bania and Kuchar 1993), or coverage of the southern galactic plane (the Kerr et al. 1986 survey). But some entries must seem inexplicable to a nonspecialist. What is the point of doing a new survey with a 36' beam (e.g. Hartmann and Burton 1993) when telescopes are available that give up to an order of magnitude more resolution? And why would anyone be interested in the Stark et al. (1992) survey with its awful resolution in both angle and velocity? This paper tries to answer these questions and provide a partial users guide to galactic H I data sets. There is a clear need for such a guide. As noted by Heiles and Wrixon (1976, in an article that contains an excellent description of some of the vicissitudes of 21 cm observing), galactic H I observers are uncommonly generous with their data and often distribute it well in advance of publication - there are many H I data sets available in digital form. But the data contain pitfalls for the unwary and cannot be used uncritically. Interested parties should consult a recent review by Dickey and Lockman (1990, hereafter DL90) that gives a detailed treatment of the topics discussed here and also covers other issues, like the use of interferometers for 21 cm research, that there is not space to include here.

2. EMISSION OR ABSORPTION?

The brightness temperature above the background continuum of an H I line from a single element of gas at a single velocity is

$$T_{b(v)} = (T_s - T_o) (1 - e^{-\tau(v)})$$

where $\tau(v)$ is the optical depth, T_s is, in most circumstances, the physical temperature of the

Workshop on Databases for Galactic Structure
A. G. Davis Philip, Bernard Hauck and Arthur
R. Upgren, eds. p. 181 - 188 © 1993 L. Davis Press

181

Felix J. Lockman
National Radio Astronomy Observatory
PO Box 2
Greenbank, West Virginia
24844

H I, and T_0 is the brightness temperature of any radio continuum emission incident on the back of the cloud. The most significant background radiation comes from continuum sources like supernova remnants, H II regions, quasars and radio galaxies. The value of T_0 contributed by a small radio source of flux density S can be approximated

TABLE 1

Selected H I Surveys

Survey	HPBW	Δv	Coverage	Noise
Weaver-Williams 1973	36'	2.1k/s	$10° \leq l \leq 250°$; $|b| \leq 10°$	0.4 K
Heiles-Habing 1974	36'	2.1	all l; $|b| > 10°$ at $\delta > -30°$	1.2
Cleary et al. 1979	48'	7.0	all l; $|b| > 10°$ at $\delta < -30°$	0.3
Westerhout-Wendlandt 1982	13'	2.0	$11° \leq l \leq 235°$; $|b| \leq 2°$	1.0
Burton-Hekkert 1986	21'	1.0	all l; $|b| \leq 20°$ at $\delta > -46°$	0.2
Kerr et al. 1986	48'	2.1	$240° \leq l \leq 350°$; $|b| \leq 10°$	0.8
Stark et al. 1992	3° x 2°	5.3	$\delta \geq -40°$	0.02
Bania-Kuchar 1993	04'	0.5	$30° < l < 61°$; $|b| \leq 0.47°$	0.5
Hartmann-Burton 1993	36'	1.0	$\delta \geq -30°$	0.08

$$T_0 = 1.4\ D_{100}^{2}\ S$$

where S is in Janskys and D_{100} is the diameter of the radio telescope in units of 100 meters. Because radio sources with $S > 1$ Jy are uncommon, and because the effective spin temperature of H I is usually several tens of Kelvins, in almost all directions $T_0 << T_s$ and emission is favored over absorption for most telescopes.

There are two exceptions. When using an interferometer, the appropriate value for D_{100} is approximately the length of the interferometer baseline, which can be many kilometers. To an interferometer even a faint continuum source can have $T_0 >> T_s$ and an H I spectrum that is in absorption[1]. Interferometers have been used to make a number of H I absorption line surveys (see DL90 and references therein). The other circumstance is self absorption. This occurs when a cool foreground cloud blocks emission from brighter background H I. While self-absorption does not produce a $T_b < 0$, it does produce a relative "dip" in an H I spectrum that has a narrow velocity and is localized in space. Measurement of the self-absorbed component provides information on the properties of the foreground cloud, though usually with large uncertainties (see DL90). Self-absorption is seen toward nearby dark clouds, and is especially widespread in low latitude spectra observed with high angular resolution (Baker and

---

[1] Knowledgeable observers will object that the situation is more complex than this; the interferometer resolves out much of the emission leaving, in most cases, absorption by default. This, and other subtleties, are discussed in DL90.

Burton 1979). Indeed, recent observations suggest that it is variation in self-absorption rather than any true emission fluctuations that causes most of the position-to-position changes in low latitude H I spectra observed at high angular resolution (Bania and Lockman 1984). The recent BU-Arecibo survey of Bania and Kuchar (1993) contains a phenomenal number of instances of self-absorption. The existence of self-absorption is a reminder that the H I opacity goes as T_s^{-1} and therefore absorption is most easily detected from cool gas. Absorption surveys select out a particular component of the ISM whose properties might not be (and usually are not) typical of the general medium. A full understanding of the H I in any direction requires both emission and absorption profiles, but the column density of H I atoms as a function of velocity can be determined fairly accurately from the emission spectrum alone. Kulkarni and Heiles (1988) give a good discussion of these issues.

3. THE H I SKY

There is no direction in the sky that is void of galactic H I, and for the reasons discussed above, the H I usually appears in emission. Observed peak brightness temperatures range from 0.5 to 120 Kelvins. General properties of the H I sky are given in DL90 and only summarized here. At low galactic latitude one sees the projection of the ensemble of H I structures. Since a typical velocity gradient due to galactic rotation is only 5 km/s/kpc along the line of sight, two clouds with an internal velocity dispersion ~ 5 km/s must be separated by more than a kpc before they appear as distinct features in an H I spectrum. The lowest latitude data are therefore used mostly for study of the large-scale distribution and morphology of the atomic ISM (e.g., Burton 1988), and less for study of individual objects, though there are prominent exceptions.

At intermediate and high galactic latitudes the H I sky is fairly smooth on angular scales $< 1°$ and seems to consist of large sheets of gas, H I shells and shell fragments like the North Polar Spur and the North Celestial Pole Ring, and a few large clouds like those in Orion, Taurus and Ophiuchus (DL90). It requires data over hundreds of square degrees to be able to study or even recognize high-latitude structures (e.g., Heiles 1983). For this work, and for recognizing the general features of galactic structure, modest angular resolution is sufficient. Indeed, the overall picture of the Galaxy's structure in H I that was developed from observations with antenna beams $> 35'$ in diameter has not been modified significantly by subsequent higher resolution observations. Individual objects, however, often cannot be understood without observations on the 1' scale (e.g., Joncas, Boulanger and Dewdney 1992), although coverage of a substantial area of the sky at this high a resolution takes a prohibitively long time, as will be discussed in a subsequent section.

4. BIGGER IS NOT BRIGHTER

I mentioned above that the peak brightness temperature of H I varies by a factor ~ 250 over the sky. To first order the variation is due merely to the cosecant(b) variation in the length of the path through the Galaxy (DL 90). Thus, in almost all cases, the galactic H I brightness changes smoothly with position on the sky. [One exception is the high velocity clouds, which have a high degree of structure on fairly small angles. But even in these, most of the gas is distributed rather smoothly (Wakker and Schwarz 1991)]. Unfortunately, the relative smoothness of H I emission means that signals are no brighter to a large telescope than to a small one. The increased number of photons collected by a large aperture is offset by the proportionally smaller area of the sky that contributes the photons. To a good approximation then, H I line intensities are independent of telescope size and thus the time needed to obtain a detection is also largely independent of telescope size. For galactic H I, big telescopes are no faster than small ones, they just give better angular resolution.

5. A MATTER OF TIME

With a state-of-the-art 21 cm receiver an H I spectrum can be measured with a noise level $\sigma(T_b) \sim 0.5\ t^{-1/2}$ where the integration time t is in seconds. Thus in one second a galactic H I emission spectrum can be measured with a signal-to-noise ratio in the range 1 - 250 depending mainly on the latitude of the observation.

For complete coverage of some part of the sky, observations must be spaced by half the half-power beam-width (i.e., by HPBW/2) in both coordinates. For a radio telescope at 21 cm, HPBW $\sim$ 10' D_{100}^{-1} and therefore the number of spectra needed per square degree is ~ 144 D_{100}^{2}, and the total number of H I spectra in a survey of the northern sky at $\delta \geq$ -40° is 4.8 x 10^6 D_{100}^{2}. At an integration time of 25 seconds per spectrum, which gives a S/N ratio of $\geq$ 5 at all positions,

$$t_{tot} = 1.2 \text{ x } 10^8\ D_{100}^{2}\ \text{seconds},$$

or 1400 D_{100}^{2} days at 24 hours observing per day.

This much observing time is not available on the largest instruments, where $D_{100} \geq 1$, so almost all surveys, even the recent ones, have been done with $D_{100} = 0.25$ class telescopes and even then the sky is usually undersampled so that the effective coverage is that of a $D_{100} = 0.17$ telescope (i.e. spectra are taken only every 0°5 in each coordinate). This fundamental limitation accounts for the persistent use of 25 meter class telescopes in galactic studies.

6. STRAY RADIATION, DYNAMIC RANGE AND CONTAMINATION

Besides detecting the signals from the desired direction on the sky, a radio telescope must also reject signals from other directions. Radio telescopes of standard design do not do the later task well, especially for galactic H I. The problem arises because radio telescopes typically have $\sim$ 80% of their response in the main beam, and another 10% in "near" sidelobes that see much the same sky as the main beam. But $\sim$ 10% of the response is directed very far from the main beam, $\sim$ 70° away in some cases, in "far" sidelobes that arise from scattering of radiation off blockage in the aperture such as the feed support legs. Since H I emission fills the sky it also fills all sidelobes and is indistinguishable from radiation that enters the receiver via the main beam. If a significant fraction of the far sidelobes lies on bright H I in the galactic plane while the main beam is directed towards a region of low H I, the observed H I spectrum can contain roughly equal amounts of "real" and "stray" radiation. Sidelobes reduce the effective dynamic range of 21 cm spectra, in some cases drastically (this is discussed in, e.g., Kalberla et al. 1980, Lockman et al. 1986; DL90).

There are two cures for stray radiation: (1) use an unblocked telescope that has little response outside the main beam, or (2) correct the data for stray radiation using deconvolution or bootstrap techniques. The only unblocked telescope that has been used recently for H I work is the AT&T Bell Labs horn-reflector. It has been used to survey the sky north of δ = -40° and produce the "cleanest" H I data set yet available (Stark et al. 1992). The absence of stray radiation in this data set more than compensates for its poor angular and spectral resolution.

Several groups taking the second approach attempt to correct observations for stray radiation retrospectively (Kalberla et al. 1980, Lockman et al. 1986, Willacy et al. 1993). This technique was used by Hartmann and Burton in their new survey discussed in Section 7.2.

For observations near the galactic plane, where the H I in the forward direction is bright while that coming in through far sidelobes is faint, contamination is relatively small, and most information derived from the low-latitude surveys is accurate. The stray radiation merely reduces the believable dynamic range of the data. This is the case for the new BU-Arecibo survey discussed in Section 7.1. For high galactic latitude observations, however, stray radiation can corrupt spectra entirely. The Green Bank Telescope, a 100 meter diameter unblocked instrument that will come into operation in a few years, will be able to measure galactic H I spectra at reasonable angular resolution ($\sim$ 9') with a minimum of stray radiation. It will be a powerful tool for the study of galactic H I. But there will continue to be a role for other telescopes in H I observations, and they must confront the issue of stray radiation.

7. TWO RECENT SURVEYS

The two most recent surveys listed in Table 1 each make substantial contributions to our knowledge of the H I sky, but differ so much in their properties that it is worthwhile discussing them separately to illustrate their features and limitations.

7.1 The Boston University-Arecibo Survey.

The 305 meter diameter radio telescope at Arecibo, PR, was used between 1986 - 1990 by Bania and Kuchar to measure 21 cm H I profiles at 4' angular resolution in more than 25,000 directions. The majority of the spectra were concentrated on a region of the inner galactic plane between about $30° \leq l \leq 60°$ and $|b| \leq 0°5$. Coverage of this area was good: spectra were taken every 4' over the entire survey region, and many sections had complete coverage at 2' spacing. Many other fields were mapped, usually centered on interesting objects: a star cluster, H II region or bi-polar flow. This is the highest angular resolution survey of the galactic plane that has been made (Kuchar and Bania 1990, Kuchar 1992, Bania and Kuchar 1993).

These data, with their high angular and velocity resolution, and good coverage of the galactic plane, are useful for studies of the galactic rotation curve, of H I associated with molecular clouds, and of the H I properties of objects like H II regions and SNRs that lie in the covered region. The spectra contain stray radiation, but because the H I emission from the galactic plane is very strong, contamination amounts to no more than a few percent.

7.2 The Leiden-Dwingeloo Survey

The 25 meter diameter radio telescope at Dwingeloo, The Netherlands, has been used in the past three years by Burton and Hartmann to survey the entire sky north of declination -30° at an angular resolution of 36'. The coverage is almost complete, with spectra spaced every 30'. The data have been corrected for stray radiation using the technique of Kalberla et al. (1980), and the noise level, 0.08 K rms, is quite good (Hartmann and Burton 1993).

The Dwingeloo survey will be invaluable for studying H I at moderate and high galactic latitudes, and especially for delineating large-scale features like H I shells. Also, the correction for stray radiation will make the survey useful for estimating total column densities, and for analysis of weak H I features in the profile wings. It will not supplant existing surveys of the galactic plane, but will be the data set of choice for the high latitude sky.

8. COLUMN DENSITY DETERMINATION

Quite often the most important thing about galactic H I is what it conceals, rather than what it reveals. This is especially true for studies of the UV or soft x-ray properties of extragalactic objects like QSOs. Absorption of energetic photons by the gas in our Galaxy can be severe, and extragalactic spectra must be corrected through an estimate of the amount of absorption, which is often simply proportional to N_H, the total number of hydrogen atoms along the line of sight. This has been an increasingly important use of H I surveys in recent years. An analysis of the uncertainties in N_H derived from 21 cm observations is developed in many places (Lockman et al. 1986, Elvis et al. 1989, Jahoda et al. 1990, DL90, issues of calibration and baseline removal are discussed extensively in Heiles and Wrixon 1976). Here the situation will be summarized in Table 2, which applies only to high latitude directions. Near the galactic plane, or in directions where the H I is sufficiently bright to require a significant opacity correction, the errors are harder to quantify and there might not even be a unique N_H that can be derived from the data. The opacity correction is always problematical and can introduce considerable uncertainty in the results (see Dickey and Benson 1982, Kulkarni and Heiles 1988, DL90).

The errors in N_H depend on the individual survey and also somewhat on the direction of interest, so Table 2 is no more than a rough guide. The values given are for spectra with good sensitivity (noise level ~ 0.1 K). Any user of H I data should take care to understand the magnitude and importance of the various errors lest the data be used inappropriately.

9. FINAL COMMENTS

The H I surveys now available require some knowledge and sensitivity to be used properly. The diverse properties of the different surveys in many ways reflect the diversity of the ISM: low-resolution highly accurate data reveal structure of the high-latitude sky; high angular resolution data reveal the structure of individual clouds. Each survey should come with its own warning about stray radiation, sensitivity, calibration, etc., in addition to the usual description of removal of instrumental spectral response and so on. Few now do.

There have been some recent, unfortunate, instances in which "armchair" astronomy, wherein research is done solely from existing data bases, has crossed the subtle line into "barstool" astronomy, wherein enthusiasm for the results derives from diminished judgement. For a medium as diverse as galactic H I, there might never be a data set sufficiently detailed, or sufficiently free of instrumental effects, that it can be used without some sober thought, experience, and caution.

TABLE 2

Errors in N_H at High Galactic Latitudes

Source	Error	Comments
Calibration	$\leq 5\%$	
Noise	few x 10^{18} cm^{-2}	
Baseline	few x 10^{18} cm^{-2}	
Stray Radiation	10^{19} - 10^{20} cm^{-2}	Extremely variable
Opacity Correction	$< 10\%$	Error small for low N_H

ACKNOWLEDGEMENTS

The National Radio Astronomy is operated by Associated Universities, Inc., under a cooperative agreement with the National Science Foundation. This paper was written while I was a visitor at the Aspen Center for Physics, and I thank them for their hospitality.

REFERENCES

Baker, P. L. and Burton, W.B. 1979   A&AS 35, 129
Bania, T. M. and Kuchar, T. A. 1993, in press
Bania, T. M. and Lockman, F. J. 1984 ApJS 54, 513
Burton, W.B. 1988, in Galactic and Extragalactic Radio Astronomy, K. Kellermann and
    G. L. Verschuur, eds., Springer, New York, p. 295
Burton, W. B. and te Lintel Hekkert, P. 1986 A&AS 65, 427
Cleary, M., Heiles, C. E. and Haslam, G. 1979 A&AS 36, 95
Dickey, J. M and Benson, J. M. 1982 AJ 87, 278
Dickey, J. M. and Lockman, F. J. 1990 Ann. Rev. Astr. Ap. 28, 215 (DL90)
Elvis, M., Lockman, F. J. and Wilkes, B. J. 1989 AJ 97, 777

Hartmann, D. and Burton, W.B. 1993, in press

Heiles, C. E. 1983 in Surveys of the Southern Galaxy, W. B. Burton and F. P. Israel, eds., Reidel, Dordrecht, p. 195

Heiles, C. E. and Habing, H. J. 1974 A&AS 14, 1

Heiles, C. E. and Wrixon, G. T. 1976 in Methods of Experimental Physics, Vol. 12C, M. L. Meeks, ed., Academic Press, New York, p. 58

Jahoda, K., Lockman, F. J. and McCammon, D. 1990 ApJ 354, 184

Joncas, G., Boulanger, F. D. and Dewdney, P. 1992 ApJ 397, 165

Kalberla, P. M. W., Mebold, U. and Reich, W. 1980 A&A 82, 275

Kerr, F. J., Bowers, P. F., Jackson, P. D. and Kerr, M. 1986 A&AS 66, 373

Kuchar, T. A. and Bania, T. M. 1990 ApJ 352, 192

Kuchar, T. A. 1992, Unpublished Ph.D. Thesis, Boston Univ.

Kulkarni, S. R. and Heiles, C. E. 1988 in Galactic and Extragalactic Radio Astronomy, K. Kellermann and G. L. Verschuur, eds., Springer, p. 95

Lockman, F. J., Jahoda, K. and McCammon, D. 1986 ApJ 302, 432

Stark, A. A., Gammie, C. F., Wilson, R. W., Bally, J., Linke, R. A., Heiles, C. and Hurwitz, M. 1992 ApJS 78, 77

Wakker, B. P. and Schwarz, U. J. 1991 A&A 250, 484

Weaver, H. and Williams, D. R. W. 1973 A&AS 8, 1

Westerhout, G. and Wendlandt, H. -U. 1982 A&AS 49, 143

Willacy, K., Pedlar, A. and Berry, D. 1993 MNRAS 261, 165

GALACTIC SCALE SURVEYS AT CENTIMETER THROUGH SUBMILLIMETER WAVELENGTHS

T. M. Bania

Boston University

The observational scope and astrophysical context of a sample of galactic scale sky surveys of the centimeter through submillimeter wavelength spectral lines emitted by the atoms, molecules and ions of cosmically abundant elements are discussed. This paper is not a comprehensive review of all such surveys. Rather it describes some recent, on-going, and planned surveys of particular importance to galactic structure studies. Details of the AST/RO (Antarctic Submillimeter Telescope and Remote Observatory) Project's survey of the distribution of atomic carbon in the Milky Way scheduled to commence at the South Pole in February 1994 are described for the first time. Galactic scale surveys made with state-of-the-art detectors operated in hostile environments resurrect the old observational strategic conundrum which pits microscopic (e.g. source catalogs) sampling against macroscopic databases that fully sample a large fraction of the sky to the resolution limit of the instrumentation. If, as will be the case for the AST/RO atomic carbon survey, the mapping rate is low, it is clear that the greatest scientific gain will be achieved by a hybrid approach. It is argued that a modern version of Kapteyn's Plan of Selected Areas needs to be established in regard to surveys of the Milky Way's interstellar medium. It should be feasible to identify a finite number of zones distributed throughout the sky which together provide a representative sampling of sources and regions that can be used to address a myriad of issues in regard to galactic structure and the astrophysics of the interstellar medium. The utility of a new Plan of Selected Areas for galactic structure studies is obvious. In particular, the on-going problem of properly comparing galactic structure databases observed from Earth's two hemispheres becomes much more tractable if there is a significant observational overlap. Multi-wavelength studies could also be completed much more quickly and efficiently.

Workshop on Databases for Galactic Structure
A. G. Davis Philip, Bernard Hauck and Arthur
R. Upgren, eds. p. 189 - 190 © 1993 L. Davis Press

T. M. Bania
Dept. of Astronomy
Boston University
Boston, Massachusetts
02215

DISCUSSION

RATNATUNGA: What plans are there to solve the problem of snow accumulation at the site?

BANIA: It only snows about an inch or two each year, but of course it never goes away! The real problem is drifting. They bulldoze drifts continuously. We think the building design will give us an operating lifetime of 10 - 15 years at minimum.

EPCHTEIN: Could you give us any information about projects for near IR surveys from the antarctic?

BANIA: I am not really competent to address this question adequately. CARA (Center for Astrophysical Research in Antarctica) is planning to do a survey in the IR near 2.8 microns wavelength. This is the SPIREX (South Pole Infrared Explorer) project. You should contact Al Harper at Yerkes Observatory for more details in this regard.

CONCLUDING REMARKS

W. F. van Altena

Yale University Observatory

We started this "Workshop on Databases for Galactic Structure" by learning that most of us deal with Databanks, i.e. the repository of data, and hope to create a Database for the community of users, which is the Databank and the software that enables one to access and manipulate the data with some ease. Based on these definitions, it unfortunately appears that we now have very few Databases in Astronomy although we have heard several speakers discuss their progress towards providing real Databases!

Next year will be the Year of the Catalogues with new editions of the Bright Star Catalogue (Warren 1993), the Catalogue of Nearby Stars (Jahreiss and Gliese 1993), the General Catalogue of Trigonometric Parallaxes (van Altena, Lee and Hoffleit 1993) and the Washington Visual Double Star Catalogue (Worley and Douglass 1993) being scheduled to appear in 1994. Sadly lacking is a new edition of the badly out-of-date Radial Velocity Catalogue and it was noted several times that a central data bank for spectroscopic information has been proposed and funding is being sought (Griffin 1993).

New proper motion catalogues were discussed by several speakers and we were made aware of important differences between the two new major proper motion compilation catalogues: the U. S. Naval Observatory's ACRS (Urban 1993); and the Astronomisches Rechen Institut's PPM (Roeser 1993). In the first case, the Astrographic Catalogue Reference Stars is designed to re-reduce the old Astrographic Catalogue which has an average epoch in the early 1900's. It therefore contains no AC positions and is designed to have an old mean epoch. On the other hand, the Positions and Proper Motions Catalogue is designed to provide the best current proper motions and positions and it therefore contains most available positions and has a more recent mean epoch. It is apparent therefore that comparisons between the two catalogues are of little value and that the choice of which catalogue to use for the reduction of plate material will depend on the epoch of the plates to be reduced. Progress with the all-sky proper motion surveys with respect to faint galaxies currently underway at Lick for the northern sky and at Yale and San Juan for the Southern Sky was also described by Platais et al. (1993). The northern part of the sky outside the Milky Way will be published soon by Lick, while the Milky Way and the Southern Sky still have a considerable way to go. We also heard that once the data from the Hipparcos satellite are published, the Tycho observations will be combined with the turn of the century re-reduced Astrographic Catalogue positions to yield proper motions for about 1,200,000 stars (Roeser 1993). After so many years of having to rely on the SAO Star Catalogue as the only major integrated source of proper motions, we are now in the happy position of having two quality proper motion catalogues in our hands and in the near future the even more accurate Hipparcos and Tycho-AC catalogues.

Workshop on Databases for Galactic Structure
A. G. Davis Philip, Bernard Hauck and Arthur
R. Upgren, eds. p. 191 - 196 © 1993 L. Davis Press

William F. van Altena
Yale University Observatory
New Haven, Connecticut
06511

Our ability to access the best current data for an individual star, i.e. a comprehensive stellar database, is being worked on by several independent groups including the CDS at Strasbourg (Crézé 1993), the SAO (Eichhorn 1993) and for Open Clusters at Lausanne (Mermilliod 1993). It is clear that enormous progress has been made in our ability to access the data in many catalogues for an individual star or class of stars. On the other hand, much remains to be done to provide an interface to the user that will be truly "friendly". We should also remember that in our efforts to make available the best data in the shortest possible time that we must not forget those without the latest technology at their fingertips. Even though we are now able to access the newest data at various data banks through electronic networks, we should continue printing new catalogues so that all astronomers will be able to work even if they do not have the latest computer equipment. As in the past, the Astronomical Data Center at the Goddard Space Flight Center continues to make catalogues available to its users in a variety of formats and has now announced the publication of a new DM Catalogue in paper format (Roman 1993). We all owe the dedicated ADC and CDS staffs, past and present, a great deal for their work in support of the astronomical community!

The subject of this Workshop was Databases for Galactic Structure and we heard from several speakers about their preparation and use of various databases to study our Galaxy (Philip and Upgren 1993, Straižys, Boyle and Philip 1993). Investigations were also described whose goals were to study: the structure towards the galactic poles (Lu 1993), and at low latitudes (Tsay et al. 1933); Field Horizontal Branch stars to probe the structure of the Galaxy (Philip and Adelman 1993); and the nature and content of Galactic Open Clusters (Casertano et al. 1993, Lattanzi 1993, Mermilliod 1993).

Astronomy has evolved into a science that not only deals with enormous distances, but now we have truly moved into the "digital age" and have available to us amounts of data that were unthinkable only a few years ago. The always valuable Schmidt Surveys and their current second epoch surveys made with the 48-inch Palomar, UK/AAO and ESO Schmidt Telescopes are now being scanned and digitized by several groups, including the Space Telescope Science Institute (Lasker 1993, Sturch et al. 1993), the Royal Observatory Edinburgh (MacGillivray 1992), Cambridge University (Irwin 1992), the University of Minnesota (Humphreys 1993), and the U. S. Naval Observatory at Flagstaff (Pier and Monet 1993). Out of these many terabytes of data will come magnitudes, colors and proper motions for more than one hundred million stars and galaxies from which we will be able to study the structure of the Galaxy and the Universe without the troublesome selection effects that often dominated earlier investigations. In addition, large multi-Terabyte digital sky surveys are being planned which use detectors to observe the sky directly in the "stare" or "drift" modes in wavelengths from the visual (Sloan: Richmond 1993) through the infra-red (DENIS: Epchtein 1993; 2MASS (Kleinmann 1993); TMGS: Hammersley et al. 1993) to the radio regions of the spectrum (Lockman 1993 and Bania 1993).

When I reflect on the quantity of data that we have recently obtained with the publication of the 20 million star HST Guide Star Catalogue (Lasker et al. 1990) and how that number seems small in the context of what is to come, I am awed at the multitude of problems that we will be able to tackle and how many of the selection biases that effected earlier investigations with less than complete coverage will be eliminated or dramatically reduced. However, I am troubled by the prospect of being inundated with data both as a producer and as a user. How do we enforce and maintain quality assurance with terabytes of data? How do we digest and understand such quantities of data? In the past we were forced to squeeze every bit of information out of the small amounts of data available, but now will we be reduced to, as one speaker commented, "Bar stool analyses" or simply skimming the top of the data? I believe that one of the major topics that must be addressed by the groups that are digitizing the sky survey plates and those that are performing digital sky surveys is how can they enforce a high level of quality assurance on their product. A thoughtful and detailed analysis of this topic by all of the associated groups will help to avoid many problems and uncertainties at a later date.

Certainly, one way of condensing the terabytes of data discussed above is to describe the data in terms of a model of the Galaxy and the distribution of galaxies. We heard several presentations (Crézé, Cohen, Casertano and Ratnatunga, Mendez et al. and Oblak et al., all 1993) on such parameterizations of the Galaxy and it is apparent that considerable progress has been made in modeling both the spatial and kinematic structure of the Galaxy. I believe that we will soon be in a position to model the structure of our Galaxy with some accuracy and begin to study the relevance of the deviations from our expectations.

This Workshop was a very important contribution to my understanding of where we stand today in the production of astronomical Databanks and Databases that can be used with some ease for investigations of galactic structure. It also highlighted many problems and gaps that need to be dealt with before we will have an adequate understanding of the structure of our Galaxy.

We should thank our hosts who have worked hard to make this Workshop possible: Bernard Hauck, who first proposed the idea of the Workshop and carried through with its scientific organization; Wulff Heintz and Harry Augenson, who were responsible for the smooth operation of the local organization; Dave Philip, who assisted in most aspects of the scientific and local organization and also organized the publication of the proceedings, and Bruce Cantor, who spent his time during the conference typing the discussions so that we could review them before returning home; to all of these people, we are grateful for the many hours of labor that they spent to make this such a productive and enjoyable meeting.

REFERENCES

DBGS stands for Workshop on Databases for Galactic Structure, A. G. D. Philip, B. Hauck and A. R. Upgren, eds., L. Davis Press, Schenectady.

Bania, T. 1993 DBGS, p. 189
Casertano, S. and Ratnatunga, K. 1993 DBGS, p. 167
Casertano, S., Ratnatunga, K. and Lattanzi, M. 1993 DBGS, p. 227
Cohen, M. 1993 DBGS, p. 131
Crézé, M. 1993 DBGS, p. 109
Eichhorn, G. 1993 DBGS, p. 45
Epchtein, N. 1993 DBGS, p. 97
Griffin, E. 1993 DBGS, p. 43
Hammersley, P., Garzón, F., Mahoney, T., Calbet, X. and Selby, M. J. 1993 DBGS, p. 103

Humphreys, R. 1993 DBGS, p. 87

Irwin, M. J. and McMahon, R. 1992 IAU Comm. 9 Working Group on Wide-field Imaging, Newsletter 2, H. T. MacGillivray, ed., Royal Observatory Edinburgh, p. 31

Jahreiss, H. and Gliese, W. 1993 DBGS, p. 53

Kleinmann, S. 1993 in UCLA Conference on Astronomy with Infrared Arrays, I. McLean, ed.

Lasker, B. M. 1993 DBGS, p. 77

Lasker, B. M., Sturch, C. R., McLean, B., Russell, J. L., Jenkner, H. and Shara, M. 1990 AJ 99, 2019

Lattanzi, M. 1993 DBGS, p. 173

Lockman, F. 1993 DBGS, p. 181

Lu, P. 1993 DBGS, p. 19

MacGillivray, H. T. 1992 IAU Comm. 9 Working Group on Wide-field Imaging, Newsletter 2, H. T. MacGillivray, ed., Royal Observatory Edinburgh, p. 34

Mendez, R. A., Platais, V. and Girard, T. M. 1993 DBGS, p. 223

Mermilliod, J.- C. 1993 DBGS, p. 27

Oblak, E., Zola, S. and Chareton, M. 1993 DBGS, p. 211

Philip, A. G. D. and Adelman, S. J. 1993 DBGS, p. 245

Philip, A. G. D. and Upgren, A. R. 1993 DBGS, p. 235

Pier, J. and Monet, D. 1993 DBGS, p. 161

Platais, I., Girard, T. M., van Altena, W. F. and Lopez, C. E. 1993 DBGS, p. 153

Richmond, M. 1993 DBGS, p. 207

Roman, N. 1993 DBGS, p. 39

Roeser, S. 1993 DBGS, p. 137

Straižys, V., Boyle, R. and Philip, A. G. D. 1993 DBGS, p. 123

Sturch, C. R., Laidler, V. G., Lasker, B. M. and Postman, M. 1993 DBGS, p. 201

Tsay, W. -S., Lu, P. K. and van Altena, W. F. 1933 DBGS, p. 241

Urban, S. 1993 DBGS, p. 145

van Altena, W. F., Lee, J. T. and Hoffleit, E. D. 1993 DBGS, p. 65

Warren, W. H. 1993 DBGS, p. 217

Worley, C. E. and Douglas, G. G. 1993 DBGS, p. 35

Poster Papers

Chairmen: Discussion Sections

Discussion 1:   Barry Lasker

Discussion 2:   Alain Omont

Discussion, Poster
        Papers: A. G. Davis Philip

STARBASE: THE AUTOMATED PLATE SCANNER CATALOG OF THE PALOMAR SKY SURVEY

R. M. Humphreys, S. C. Odewahn, G. S. Aldering and P. M. Thurmes

University of Minnesota

1. THE APS SCANNING PROJECT

The Automated Plate Scanner (APS) is a unique high-speed "flying spot" laser scanner. Originally designed as a proper motion measuring machine, the APS has been redeveloped using modern electronics and computers to produce a very flexible scientific instrument. The APS uses a rotating prism in combination with a beam splitter to produce a pair of spots which scan across a pair of plates. Because of this unique design, the APS is able to scan and digitize two 14 inch square Schmidt plates simultaneously in about six hours. Using a series of low resolution background scans and high resolution "forward" scans, the APS utilizes a thresholding mode to detect objects which lie above some user specified plate transmittance level. Pixel data are collected for all regions of the scan detected above this threshold level. All images are reconstructed and parameterized during the scanning process by software running on both a SUN Sparc I and a Sparc 10. A Silicon Graphics R-4D220S is also available for post-reduction processing.

The APS group has just completed scanning the high latitude (b $>$ $|20°|$) glass copies of the blue and red (O and E) plates in the Palomar Observatory - National Geographic Sky Survey (POSS I). The scanned fields are shown in Fig. 1. All of the detected images are being cataloged with positions, isophotal shapes, isophotal diameters, magnitudes and colors. The stellar and non-stellar objects are separated using a set of neural network image classifiers (Odewahn et al. 1992, 1993). Densitometric data will be collected for all extended objects. We estimate that there will be nearly a billion stellar objects and image data for several million galaxies when the project is completed. The final goal of this work is a database, including the cataloged objects and the pixel data for the non-stellar images, which will be made available to the astronomical community on-line via Internet. The APS Catalog of the POSS I will complement other existing or planned sky-survey size catalogs such as the Guide Star Catalog (Lasker et al. 1990). We summarize the properties of our Catalog in Table I below. The scanning project and the Catalog are described in detail in Pennington et al. (1993).

2. A SUMMARY OF STARBASE

STARBASE is a custom database engine which is being uniquely tuned to the requirements of the APS Catalog. The package was designed and written at the University of Minnesota by E. B. Stockwell under the supervision of R. L. Pennington. STARBASE has been written to provide fast access to the hundreds of millions of images collected by the APS. This system provides an initial reduction mode for parameterizing APS images and classifying image types

Workshop on Databases for Galactic Structure
A. G. Davis Philip, Bernard Hauck and Arthur
R. Upgren, eds. p. 197 - 199 © 1993 L. Davis Press

Roberta M. Humphreys
Dept. of Astronomy
University of Minnesota
Minneapolis, Minnesota
55455

TABLE 1

APS POSS I Catalog Characteristics

Stars Catalog:
   10^9 records per color
   Magnitude range: 12-22 in O, 12-20.7 in E
   Matched (O and E) images only
   Astrometry accurate to 0.4 arcseconds
   Photometry accurate to 0.1 to 0.2 mag

Galaxies Catalog:
   $Nx10^6$ records per color
   Integrated isophotal magnitudes: 12-20 in O, 12-19.5 in E
   Match (O and E) images only
   Photometry accurate to 0.3 to 0.5 mag

using a novel set of neural network image classifiers. A second analysis mode, which will be that commonly used by the general user, provides for searches of the database which may be constrained by any combination of physical and positional parameters.

Each color in the survey will comprise a database of ~ 200 GB. Searching this by brute force would require prohibitive amounts of time. Search times have been greatly reduced by producing hash lists for each searchable field and two-dimensional search trees for the positional information. Search times are generally seconds or minutes for reasonably sized searches.

In addition to speed, the three major requirements that our DBMS had to fulfill are portability, minimized storage requirements, and the ability to accommodate the full set of plates as a single entity. It is portable to any 32 bit integer machine and is written to conform to the ANSI C standard. Each record in the database (1 record per object per color) is bit-packed into 45 bytes. This contains all of the parameterized information form the APS for an image, transmittance information, indices to the same object on other plates, and a pointer to an extended field that may contain other information (references to other catalogs, etc.). An image which has been classified as extended by the neural network classifier will also include pixel information in the extended field.

STARBASE may be used interactively in two different ways. First, the user may enter direct database access commands to produce ASCII tables of data for records that meet the user imposed search criteria. In the second mode, the data may be accessed and manipulated through display and graphics programs which present the data visually. In addition to fast data retrieval, the system provides a graphical interface for efficient display of scatter plots or histograms of the collected data. Using these X windows-based graphics tools, the astronomer is able to get a quick look at the characteristics of the selected data before generating the final data file(s) to be returned to the home institution by ftp. Finally, STARBASE will have a flexible programmable interface which allows other programs to access information in the database.

This will allow users to write applications suited to their particular needs to process APS data.

3. REFERENCES

Lasker, B. M., Sturch, C. R., McLean, B. J., Russell, J., Jenkner, H., Shara, M. M. 1990 AJ 99, 2019

Odewahn, S. C., Humphreys, R. M., Aldering, G. and Thurmes, P. 1993 PASP, in press

Odewahn, S. C., Stockwell, E. B., Pennington, R. L., Humphreys, R. M. and Zumach, W. A. 1992 AJ 103, 318

Pennington, R. L., Humphreys, R. M., Odewahn, S. C., Zumach, W. and Thurmes, P. M. 1993 PASP, in press

DISCUSSION

PHILIP: Have you a proposed date by which the data will all be in the online database to Internet?

HUMPHREYS: Initially we will go on line with a demonstration database on ADS with limited access. This will be for testing the interface with ADS and StarBase. After a few weeks of testing this summer we expect that Starbase should be generally available to the community by early this Fall.

ROMAN: Would we now be more likely to compare catalogs by position rather than to use optical blinking?

HUMPHREYS: I think catalogs and databases will be cross referenced by position.

SKY SURVEY PROGRAMS AT ST ScI: DIGITIZATION AND DISTRIBUTION

C. R. Sturch, V. G. Laidler

Computer Sciences Corporation at Space
Telescope Science Institute

G. R. Greene, B. M. Lasker and M. Postman

Space Telescope Science Institute

1. SCANNING

Scanning of second generation Schmidt sky survey plates is well under way at ST ScI. The goals are the following: to support HST operations by providing deep planning material heretofore unavailable in the north and determining the proper motions of HST guide stars; to perform a number of research programs in cosmology and stellar statistics; and to provide community access to scan data. The northern program includes all 894 fields of the second Palomar sky survey (POSS-II; Reid et al. 1991) in three bands: blue (IIIa-J), red (IIIa-F) and near-IR (IV-N). The southern program consists of high latitude Second Epoch Southern (SES; Lasker and Cannon 1989) red (IIIa-F) plates.

Digitization of sky surveys for construction of the Guide Star Catalog (GSC-I, cf. Lasker et al. 1990) was performed on two modified PDS scanning microdensitometers. Driven by the requirements of HST launch, we used a 25 um sampling interval for this early work. However, the scanning schedule for the second generation surveys is less constrained, and a more appropriate sampling interval for modern Schmidt plates, 15 μm, has been adopted (Laidler et al. 1992). The finer sampling results in 1.06 Gbytes of data per image and, with the original microdensitometer configuration, a much longer scan time. Because a significant increase in throughput is essential to completing the scanning soon after the new surveys are available, a microdensitometer upgrade program is now in progress.

The first phase of the upgrade program, the replacement of the original (1970s vintage) servo system with a modern digital one (HP 5507, incorporating the original metrology laser elements; cf., Kinsey, Denman, and Evzerov, 1984), has been completed on one microdensitometer, renamed the Guide star Automatic Measuring MAchine (GAMMA), and is in progress on the second machine. The servo upgrade is accompanied by a replacement of the data system and other computer interfaces by a SCSI/CAMAC system operated from a VAXstation. In the new configuration the coupling between the control computer and the servo system is sufficiently flexible that a software-diagnostic approach is now possible; this greatly simplifies the required maintenance activities. Operationally, an immediate benefit of the enhancements is the elimination of "image chopping", a scan defect due to the inability of the old Y servo to track high frequency irregularities on the Y-defining surface. The problem

Workshop on Databases for Galactic Structure
A. G. Davis Philip, Bernard Hauck and Arthur
R. Upgren, eds. p. 201 - 205 © 1993 L. Davis Press

Conrad R. Sturch
Computer Sciences Corp.
Space Telescope Science Institute
3700 San Martin Drive
Baltimore, Maryland
21218

had been controlled by limiting scanning speeds in X to 50-75 mm/sec; now the new Y servo is capable of tracking the irregularities at much higher scan speeds.

The final phase of the enhancements consists of a rework of the optical system. The new illumination source will be a red HeNe laser, brought to a very clean Gaussian spot (10 μ full width at $e^{-1/2}$) by conventional Gaussian optics. This light source, unlike the older thermal one, is sufficiently bright to support the short pixel times required for high throughput (eg., one scan in 12 hours corresponds to 60 μs per pixel). The laser illumination is also conducive to easy beam steering with an acousto-optic deflector; this generalization will be added, following the concepts of Kibblewhite (1984), to achieve a multi-channel capability. As of this writing, experimental single channel laser scanning has been successfully done and the full implementation is scheduled for the end of 1993.

2. COMPRESSION

The ST ScI is using low-loss image compression, based on the H-transform (White et al. 1992), to publish its archived digital data on CD-ROMs. The initial surveys to be compressed are the southern hemisphere UK SERC (J band) survey with the equatorial extension (894 plates; epoch 1975-1984) and the northern hemisphere National Geographic Society - Palomar Observatory (E band) survey (583 plates; epoch 1950-56). The POSS E survey was chosen because it is currently the only complete northern survey which has a depth approaching that of the SERC-J survey in the south.

Two versions are being produced: one at a compression factor of 10 which is virtually indistinguishable from the original data (cf., Fig. 1), and one at a compression factor of about 90 which will be provided to the educational and amateur communities. For the 10 x compression, it is expected that the SERC-J data will fit on 60 CD-ROMs and the NGS-POSS-E data will fit on 40 CD-ROMs. These numbers are based on current storage capacity limits of about 600 MBytes per disk.

Digitization defects such as line-shearing and the chopping described above are measured and evaluated for every image. Photometric and astrometric stability after compression are also monitored. The results are automatically logged into a database for reference and strategic analysis. Compression of the SERC-J survey was started first because extensive quality assurance had been performed on these data during construction of the GSC. Compression of these scans is essentially complete, and 98.5% of the SRC compressed scans have passed all quality assurance tests. Of the remainder, at least half can probably be "repaired" via Fourier processing techniques. A few plates may require rescanning. Premastering operations of the southern survey will begin this summer with release of the data expected later this year.

The POSS E scans have not undergone extensive quality assurance prior to this project because they were not used in GSC development. Initial compression of approximately 15 POSS scans has been performed to understand better the image quality concerns these survey data require. Full production of compressed POSS E scans will begin this summer.

The compressed data will be accompanied initially by the astrometric calibrations of the individual plates used in the GSC for the south and preliminary calibrations for the north. The

Fig. 1. Original images and restored images compressed at ratios of 10 x and 100 x of the globular cluster M 71 and Abell Cluster 1060. The 10 x image is nearly indistinguishable from the original while the decrease in resolution and limiting magnitude is apparent in the 100 x image.

astrometric recalibration planned for GSC 1.2 based on the concepts in Taff, Lattanzi and Bucciarelli (1990) will be included in a later release. No photometric calibration will be available with the initial compressed data sets; however, V- and R-band CCD observations for extending the Guide Star Photometric Catalog (Lasker and Sturch, et al. 1988) to V = 18 or V = 21 have been obtained (Postman et al. 1992). These observations will be incorporated into the survey calibration database to be issued in 1995. Currently, observations have been completed to V = 18 for approximately 75% of the the northern sequences and about 25% of the southern.

A marketing and distribution plan is being established over the next few months. Full responsibility for the management of the distribution and sale of the compressed sky survey CD-ROMs is expected to be contracted to an established scientific/educational publishing firm.

ACKNOWLEDGEMENTS

The authors gratefully acknowledge the support provided by John Bedke, Robert Brucato, Bob Denman, George Djorgovski, Jesse Doggett, Anatoly Evzerov, Mario Lattanzi, Brian McLean, Michael Meakes, Flavio Mendez, Jim Phillips, Knute Ray, Daniel Rehner, Neill Reid, Richard Rossi, Nicolas Weir, and Richard White. Riccardo Giacconi, Russell Cannon, Harry Feinstein, and Gerry Neugebauer are especially thanked for their active encouragement in establishing the arrangements underlying these programs. ST ScI is operated by AURA, Inc., under contract to NASA; image compression algorithms at ST ScI were developed under NASA grant NAGW 2166.

REFERENCES

Kibblewhite, E. 1984 in Astronomical Microdensitometry, D. A. Klinglesmith, ed., NASA Conf. Proc. No. 2317, p. 333

Kinsey, J. H., Denman, R. J. and Evzerov, A. 1984 Rev. Sci. Instrum. 55, 1960

Laidler, V. G., Lasker, B. M. and Postman, M. 1992 in Digitised Optical Sky Surveys, H. T. MacGillivray and E. B. Thompson, eds., Kluwer Academic Pub., Dordrecht, p. 95

Lasker, B. M. and Cannon, R. D. 1989 CDS Bull., No. 37, p. 13

Lasker, B. M., Sturch, C. R., Lopez, C., Mallama, A. D., McLaughlin, S. F., Russell, J. L., Wisniewski, W. Z., Gillespie, B. A., Jenkner, H., Siciliano, E. D., Kenny, D., Baumert, J. H., Goldberg, A. M., Henry, G. W., Kemper, E. and Siegel, M. J. 1988 ApJS 68, 1

Lasker, B. M., Sturch, C. R., McLean, B. J., Russell, J. L., Jenkner, H. and Shara, M. M. 1990 AJ 99, 2019

Postman, M., Siciliano, L., Shara, M., Rehner, D., Brosch, N., Sturch, C., Bucciarelli, B., and Lopez, C. 1992 in Digitised Optical Sky Surveys, H. T. MacGillivray and E. B. Thompson, eds., Kluwer Academic Pub., Dordrecht, p. 61

Reid, I. N., Brewer, C., Brucato, R. J., McKinley, W. R., Maury, A., Mendenhall, D., Mould, J. R., Mueller, J., Neugebauer, G., Phinney, J., Sargent, W. L. W., Schombert, J. and Thicksten, R. 1991 PASP 103, 661

Taff, L. G., Lattanzi, M. G. and Bucciarelli, B. 1990 ApJ 358, 359

White, R. L., Postman, M. and Lattanzi, M. 1992 in Digitised Optical Sky Surveys, H. T. MacGillivray and E. B. Thompson, eds., Kluwer Academic Pub., Dordrecht, p. 167

DISCUSSION

PHILIP: At some future time will it be possible to "blink" first and second epoch plates on line or would someone just transfer the data for each plate to his computer and blink there?

LASKER: My recomendation would be to down-load the images or catalogs from the host site and do the manipulation at the home institution. I also note that, with appropriate attention to calibration details, it should be easy to consistently operate on data from several sources.

LU: Do you have any publication or announcement of already scanned regions and fields and what is planned to be scanned later?

LASKER: We are planning to make this list, and most likely other data, network-accessible to the community in the near future.

RATNATUNGA: You said that the ten times compressed CD will be available for the southern sky this year. Could you comment on the smaller set of 100 times compressed CDs?

LASKER: In principle, this can be available at the same time; however, the details must await the selection of the distributor.

THE SLOAN DIGITAL SKY SURVEY

Michael Richmond

Princeton University

1. INTRODUCTION

The Sloan Digital Sky Survey, a project involving Princeton University, the University of Chicago, Fermilab, Johns Hopkins University, the University of Washington and a group of collaborators in Japan, will be the first large-scale survey of the sky with CCDs instead of photographic plates. Our ultimate goal is to create a three-dimensional map of the galaxy distribution out to a redshift of about $z = 0.2$, but the huge amount of information gathered along the way should provide the several camps of astronomers (those who study the Milky Way, other galaxies and even the solar system!) with the source for years of analysis.

We are currently building a 2.5-meter alt-az reflector, which will be dedicated completely to the survey. It will be located at Apache Point on Sacramento Peak in New Mexico, not far from the Sacramento Peak Station of the National Solar Observatory. At an elevation of about 2800 meters, the site has very good seeing (median of about 0.8 arcseconds) and a dark sky (21.9 magnitudes per square arcsecond in V). Although it suffers from the same monsoon season as Kitt Peak, the site manages to average about seven clear nights out of every ten over the course of a year. In addition to the large telescope, we will have a smaller instrument of about 0.75-meter aperture on location; it will perform all-sky photometry every night, as a check on the weather conditions and in order to derive the calibration constants needed to reduce data from the 2.5-meter telescope.

Starting in 1995, and continuing for five years, our survey will cover roughly one-quarter of the entire sky, centered on the North Galactic Pole. Within this area, we will examine every section of the sky twice. In the first pass, we will take images through five different filters; in the second, we will target all the "interesting" objects found in the first pass for spectroscopy. Let me discuss each method in turn.

First, the imaging: the 2.5-meter telescope will have optics that produce seeing-limited images over a field roughly three degrees in diameter. The camera consists of an array of thirty 2048 x 2048 CCDs which cover this area, arranged in six columns; each column contains a row of five CCDs, one for each of the filters in our system. The camera will operate in drift-scan mode: the telescope will track the sky at a slightly non-sidereal rate, so that stars drift down a column in the array, spending about fifty-five seconds on each CCD. There are gaps between each column of chips, so after scanning for several hours, the telescope will shift over about ten arcminutes (the size of the gap) and scan back over the same region, getting all the objects missed the first time. We expect to detect and measure the properties of point sources down to about twenty-second magnitude, with a somewhat brighter limit for extended objects.

Workshop on Databases for Galactic Structure
A. G. Davis Philip, Bernard Hauck and Arthur
R. Upgren, eds. p. 207 - 209 © 1993 L. Davis Press

Michael Richmond
Astrophysics Dept.
Princeton University
Princeton, New Jersey
08554

A brief word about the filter system: we have chosen to adopt a new set of passbands, in order to minimize interference from atmospheric emission lines and also to yield sensitive diagnostic colors for distinguishing stars from quasars. Our set has the following approximate characteristics (units are Angstroms):

TABLE 1

Filter Charactersitics

Filter	Effective Wavelength	Width
u	3510	0800
g	4730	1300
r	6270	1400
i	7690	1500
z	9250	1700

Second, the spectroscopy: after we have taken a picture of each section of the sky, we will reduce the image and analyze each object, measuring its position, brightness, size, etc., and classifying it as a star, galaxy or other object. Several weeks later, during dark time, we will go back to the area and take spectra of all the bright galaxies (down to about eighteenth mgnitude or so), any quasar candidates, and any other objects judged to be "interesting." The fiber-fed spectrograph design allows us to obtain several hundred spectra per three-degree field, and we can make several observations of very crowded regions. Each spectrum will range from about 4000 Å to 9000 Å, with a resolution of about 3 Å. Each field will receive three exposures, yielding a total exposure time of one hour. We expect to achieve signal-to-noise ratios above ten for point sources brighter than twentieth magnitude.

2. WHAT ABOUT GALACTIC STRUCTURE?

Although the primary goal of our project is to measure the properties of a complete sample of nearby galaxies, it will produce a large amount of information useful for studies of our own Milky Way. We will create a catalog of several tens of millions of stars in the region around the North Galactic Pole, down to a magnitude limit of ~ 23; for each star, we will list its brightness in five colors, to an accuracy of at least 0.1 magnitude, and its position (tied to the Hipparchos coordinate system), which we hope to determine to about 0.05 arcsecond.

In addition to the simple catalog, we will produce an enormous database containing every datum of information gathered during the course of the survey, to which the astronomical community will have some kind of remote access. We are designing the database so that it will be very flexible and allow users to ask sophisticated questions. Because our survey will contain many overlap regions, separated in time by lengths varying from hours to years, careful searches through the database for multiple observations of the same star may produce a large number of new variables. Moreover, since the survey is scheduled to take about five years, there is the possibility of measuring proper motions (of order fifty milliarcseconds per year or larger) for those stars observed over long intervals. It may also be possible to discover new clusters or associations of stars, or to define new classes of stars based on their colors. A very useful by-product of the survey will be a map of the interstellar extinction within our galaxy, much more detailed than any currently existing.

We plan to make the products of our survey freely available to any interested astronomers; I hope that in five to seven years, when the first results begin to appear, that some of you will take us up on the offer!

A PHOTOMETRIC DATABASE ON STELLAR SYSTEMS

E. Oblak

Observatoire de Besançon

S. Zola

Cracow Observatory, Jagiellonian University

M. Chareton

Observatoire de Strasbourg

ABSTRACT: We carry out photometric observations of close visual stellar systems taken from the Hipparcos Input Catalogue. We compile a general photometric database of stellar systems.

1. INTRODUCTION

The importance of studies of visual double stars does not reside only in the traditional determination of stellar masses in orbital pairs - however fundamental these may be - but also in determination of the characteristics and the frequency of double stars in different stellar populations and evolution stages. The usefulness of photometry of visual binaries is especially obvious in applications concerning, for example, luminosity calibrations, the mass-luminosity relationship, mass-ratio determination from differential magnitudes, age and evolution determination.

The technical difficulties of observing two images in close proximity one to another are especially pronounced in carrying out conventional photoelectric photometry for the remaining class of double stars with separations in the intermediate range. This class of objects has therefore largely been neglected in previous photometric programs. When available, global photometry in combination with visual or photographic estimates of the magnitude differences are not sufficiently precise to match present requirements and the accuracy achieved in other techniques (Fig. 1).

With the introduction of CCD detectors, it appears now possible to obtain accurate photometric data also for each of the component of close visual double stars with angular separations between 1 and 12".

2. THE OBSERVATIONAL PROGRAM

The aim of a European network of laboratories (Oblak et al. 1992), is to contribute to a systematic and unbiased photometric survey of components of double and multiple stars. The

Workshop on Databases for Galactic Structure
A. G. Davis Philip, Bernard Hauck and Arthur
R. Upgren, eds. p. 211 - 214 © 1993 L. Davis Press

E. Oblak
Observatoire de Besançon
F-25044 Besançon Cedex
France

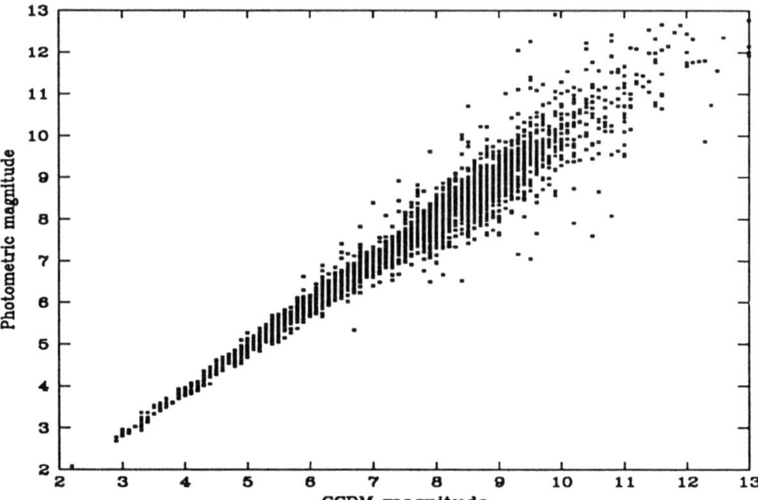

Fig. 1. Comparison between CCDM and photometric magnitudes.

available photometric data, obtained at ground-based observatories or in space programs, are mostly concerned with the global system. To obtain the relevant astrophysical information we need the individual photometric data on all the components of a double or multiple system. A program of systematic and homogeneous acquisition of precise color indices for several thousands of components of double and multiple systems has been defined in both hemispheres, with the following priorities:

1) to supplement the Hipparcos magnitudes by astrophysically significant color indices providing the additional physical parameters such as temperature and gravity;

2) to complete the photometric information for the components not included in the Input Catalogue.

Our principal scientific objective is to provide the missing photometric data needed to supplement the high quality and extensive astrometric and photometric data on known and newly detected double stars to study adequately the mechanisms of formation and evolution of double and multiple star systems.

A comprehensive catalogue of physical pairs - from the very close to the very wide ones - with a maximum number of astrometric, spectroscopic and photometric good quality data would be a highly valuable astrophysical tool because it would contain several clues concerning the star formation and consequently also the structure and evolution of our galaxy.

The selection of the program stars was made amongst 11,434 double systems, 1960 triple systems, 536 quadruple systems and 237 multiple systems from Annex 1 of the Hipparcos Input Catalogue, containing objects down to a distance of 500 pc (Turon et al. 1992).

In view of repeated photometric campaigns distributed over both hemispheres the observational program consists of a northern (declination of the primary component > -10) and a southern sample ($\delta < + 10$). The overlapping zone in declination for $-10 < \delta < +10$ is only once observed according to feasibility (Oblak and Lampens, 1992).

Since both classical and CCD photometry are considered, the selection further includes splitting according to angular separation and differential magnitude. For separations larger than 12", individual magnitudes and colors are easily obtained from classical photometry with small telescopes. For separations smaller than 15", CCD photometry is more efficient and adequate as long as the difference in magnitude is smaller than three. Astrometric information is obtained as a by-product. The overlapping range in separation has been considered for calibration purposes between two very different techniques. Intersection with the "Catalogue Photométrique des Systèmes Doubles et Multiples" (CPSDM; Oblak, 1988) allowed us to identify those sytems lacking individual photometric information.

Observations have already started in various observatories located in both hemispheres: Calar Alto, Jungfraujoch, Haute-Provence and La Palma for the northern part and La Silla for the southern part (ESO key program, Oblak et al. 1992).

3. REDUCTION OF THE OBSERVATIONS

While the conventional photometry will be reduced in a standard way, a preliminary reduction only can be done with the known packages for the photometric reduction of CCD frames. Crowded-field-photometry routines cannot be applied without modifications to most of the obtained frames, since no reference star images are available. Therefore, individual profile fittings are necessary. Accurate profiles are needed to separate the closest pairs in the program.

Specific reduction programs for the CCD observations of close double stars have been developed, mainly at Brussels by Drs. Sinachopoulos (1992) and Cuypers (to be published). A row and a column projection is applied to each CCD frame and a Franz's profile on the data according to the least squares method support by an expert system routine. The comparison of some of the CCD data with (preliminary) Hipparcos measurements shows a reasonable agreement for the specific double star parameters.

Differential magnitudes and relative positions will be obtained with high precision. Efforts are also made to write and to test additional software for extraction of individual magnitudes and color indices with an accuracy at the 1% level.

Standard star measurements allow us to correct for extinction and to transform the magnitudes to the standard VRI system. Extinction coefficients derived from photometric observations at nearby telescopes will be used when available. Finally, all observations will be processed as homogeneously as possible with a well-defined adopted "standard" procedure.

4. THE DATABASE ON PHTOMETRIC MEASUREMENTS OF STELLAR SYSTEMS

The new photometric database is compiled upon most widely used photometric systems, i.e. UBV Johnson-Morgan, Strömgren-Crawford and Geneva systems, separately for double and stellar systems with more components. At present, the database contains 13,893 entries for A or B component and global (A + B) photometry. Among these, there are 1419 measurements of both A and B components. Examination of these data yields to the conclusion that there is a selection effect visible as a lack of data for small separations and large magnitude differences.

5. FORTHCOMING EXTENSIONS

This photometric database is being systematically completed with the information from different catalogs of stellar systems (Couteau 1992, Worley and Douglas 1984) and their new versions. By arrangement with ESA, we will also be able to include the new double stars with separations greater than one arcsecond detected in the Hipparcos mission. A parallel work consists of examining and correcting the data on stellar systems of the "Centre de Données Astronomiques de Strasbourg".

ACKNOWLEDGEMENTS

This research has been supported by the Commission of the European Communities grant No CIPA-CT-93-0610, which we gratefully acknowledge.

REFERENCES

Couteau, P. 1992 Catalogue des Etoiles Doubles Couteau, Centre de Données Astronomiques de Strasbourg, Strasbourg

Dommanget, J. 1989 in Star Catalogues: A Centennial Tribute to A. N. Vyssotsky, A. G. D. Philip and A. R. Upgren, eds., L. Davis Press, Schenectady, p. 77

Oblak, E. 1988 Astrophys. Space Sci. 142, 31

Oblak, E. 1992 in IAU Colloquium No. 135, Complementary Approaches to Double and Multiple Star Research, H. McAlister and W. Hartkopf, eds., ASP Conf. Ser., Vol. 32, 454

Oblak, E., Argue, A. N., Brosche, P., Cuypers, J., Dommanget, Duquennoy, A., J., Froeschle, M., Grenon, M., Halbwachs, J. L., Jasniewicz.G., Lampens, P., Martin, E., Mermilliod, J. C., Mignard, F., Sinachopoulos, D., Seggwiss, W. and Van Desel, E. 1992 The Messenger 69, 14

Oblak, E., Argue, A. N., Brosche, P., Cuypers, J., Dommanget, Duquennoy, A., J., Froeschle, M., Grenon, M., Halbwachs, J. L., Jasniewicz.G., Lampens, P., Martin, E., Mermilliod, J. C., Mignard, F., Sinachopoulos, D., Seggwiss, W. and Van Desel, E. 1992 in IAU Colloquium No. 136, Stellar Photometry - Current Techniques and Future Developments. I. Elliot and C. J. Butler, eds, in press

Oblak, E., Lampens, P. 1992 in IAU Colloquium No. 135, Complementary Approaches to Double and Multiple Star Research, H. McAlister and W. Hartkopf, ASP Conf. Ser., Vol. 32, 339

Sinachopoulos, D. 1992 in IAU COlloquium No. 135, Complementary Approaches to Double and Multiple Star Research, H. McAlister and W. Hartkopf, ASP Conf. Ser., Vol. 32, 352

Turon, C. et al. 1992 The Hipparcos Input Catalogue, ESA SP1136, Vol.6

Worley, C. E. and Douglas, G. G. 1984 The Washington Double Star Catalog

ASTROMETRIC CATALOGUES: SUMMARY STATISTICS AND BIBLIOGRAPHIES

W. F. van Altena, I. Platais

Yale University Observatory

K. M. Cudworth

Yerkes Observatory, University of Chicago

B. M. Lasker

Space Telescope Science Institute

J. L. Russell and K. Johnston

Naval Research Laboratory

ABSTRACT: Summary statistics and bibliographies are presented in Appendix I for a variety of optical and radio astrometric catalogues, major Schmidt telescope surveys and their digitization and proper motion membership, and space velocity studies in open and globular clusters and dwarf spheroidal galaxies. Comments, corrections and additions to the data in these tables are solicited.

1. A BRIEF DESCRIPTION OF THE CONTENTS OF THE VARIOUS TABLES

The location and identification of sources of astrometric information for various astronomical objects is an often posed question for specialists in astrometry. When van Altena was asked to prepare some astrometric material for inclusion in the new edition of Astrophysical Quantities, which is being edited by Cox (in preparation), he consulted with several individuals and decided that the most useful astrometric information would be in the form of references to specific catalogues where positions, proper motions, parallaxes and other qualities could be located. Due to the scope of the information that should be presented, co-authors were conscripted who had a detailed knowledge of specific sub-disciplines. In the following, tables and their references have been provided by the identified co-authors and organized by van Altena. In particular, the sections on "Catalogues of Positions and Proper Motions", "Future Catalogues and Surveys in Progress", "Modern Relative Proper Motion Surveys", "Stellar Compilation Catalogues" and the "Bibliography" were prepared by van Altena; "Major Schmidt Surveys and Their Digitization" and "Digitization of Schmidt Plates" were prepared in collaboration with Lasker; "Astrometric Data for Open Clusters" was prepared by Platais; "Astrometric References for Globular Clusters" was prepared by Cudworth; and "Radio Astrometry References" was prepared by Russell and Johnston.

Workshop on Databases for Galactic Structure
A. G. Davis Philip, Bernard Hauck and Arthur
R. Upgren, eds. p. 215 - 216 © 1993 L. Davis Press

William F. van Altena
Yale University Observatory
New Haven, Connecticut
06511

Due to the rather extensive nature of the tables of summary data and references this material may be found in Appendix I of this volume. The summary data provided includes; the number of stars in each catalogue, the sky coverage (i.e. declination range), central epoch of the positions, magnitude limit, error of the position at the central epoch and at 1995, error of the proper motions and a brief description of the catalogue. The section on Schmidt surveys includes information on the survey filter/emulsion combination, declination range and centers of the plates, corrector used, number of plates, epoch of the plates, location of the plate archive and the distributor of the survey, and the magnitude limit. The Digitization of Schmidt Plates includes the name of the group that is digitizing the surveys, the various surveys being digitized, the pixel size, status of the project and plans for distribution. The section on Open Clusters includes data for each cluster giving the year of publication of the astrometric study, field size studied, limiting magnitude, number of stars, number of plates, plate scale of the telescope, epoch difference of the plates, accuracy of the proper motions and the maximum probability of membership. The section on Globular Clusters and Dwarf Spheroidal Galaxies includes the number of stars studied, the magnitude limit and an indication if the space velocity was determined. Finally, the Radio Astrometry section includes the number of radio sources in the catalogue, the average error of the positions, a description of the type of catalogue, the wavelength passband, sources of the zero point of the right ascensions, and the declination limits.

ASTROMETRIC DATA IN THE BRIGHT STAR CATALOGUE (BSC)

Wayne H. Warren Jr.

Hughes STX Corporation
Laboratory for Astronomy and Solar Physics
NASA Goddard Space Flight Center

Dorrit Hoffleit

Yale University

ABSTRACT: Positions and proper motions of the highest possible accuracy are now being added to the BSC (5th edition) to replace information originally based on transformations of SAO data to J2000.0. The sources of the updated data are several new astrometric catalogs that have recently been compiled at the Astronomisches Rechen-Institut in Heidelberg and at the U. S. Naval Observatory in Washington. The procedures for incorporating the data using a hierarchical scheme will be described, as will techniques for handling close double stars, not surprisingly the most troublesome objects with which to deal.

1. INTRODUCTION

The present (published) fourth edition of the Bright Star Catalogue (BSC4, Hoffleit 1982) contains positions for equinox and epoch B1900 and B2000, and proper motions for B1950.0F while the preliminary fifth edition (BSC5, Hoffleit and Warren 1992) includes positions and proper motions for J2000.0 that have been transformed from B1950.0 data, the primary source in both cases being the Smithsonian Astrophysical Observatory Star Catalog (SAO Staff 1966). The recent completion of several new catalogs of astrometric data for equinox and epoch J2000.0 has made the replacement of the SAO data with newer, more accurate positions and proper motions possible. Although the original positional data were included primarily for reference purposes, it has come to our attention that BSC positions and proper motions are often used to set large telescopes and space platforms for observational purposes. Thus, to avoid possible future problems caused by decreasing accuracies of the older SAO data, we have been encouraged by a number of colleagues to replace those data with improved information from the newer catalogs. While we intend to retain the B1900 positions for reference purposes, the J2000.0 positions and proper motions are being incorporated at full astrometric precision ($0^{s}001$ and $0''01$ in right ascension and declination (α, δ) and 0.0001s and $0''001$ in μ_{α} and μ_{δ}, respectively). The unit of proper motion in α, which is now seconds of arc, will be replaced by seconds of time. This paper describes the procedure being used to replace the older J2000.0 astrometric data with those from a number of selected catalogs on the system of the FK5.

2. THE SOURCE DATA

The J2000.0 positions and proper motions in the preliminary BSC5 are rounded to 0.1s

Workshop on Databases for Galactic Structure
A. G. Davis Philip, Bernard Hauck and Arthur
R. Upgren, eds. p. 217 - 221 © 1993 L. Davis Press

217

Wayne H. Warren Jr.
Lab. for Astronomy and Solar Physics
NASA/Goddard Space Flight Center
Greenbelt, Maryland
20771

and 1" in α and δ, respectively, while μ_α and μ_δ are reported to 0"001. The quantity μ_α, as taken from the source catalogs, will not be converted to seconds of arc, since the unit of time is more compatible with the right ascensions given. To obtain the highest quality astrometric data, a hierarchical list of source catalogs was established by examining quoted mean errors of positions and proper motions for all source catalogs and by consultation with several astrometrists, primarily the compilers of the new catalogs. Table 1 gives the information found for all of the source catalogs in the order lowest to highest priority.

TABLE 1

Astrometric Source Catalog Information

Catalog	N	Epoch $<n>$	Pos Error		PM Error			
			α	δ	α	δ	μ_α	μ_δ
SAO South	126000	2.0	1923.4	1926.5	0.20	0.20	1.40	1.40
SAO North	133000	2.0	1927.2	1922.8	0.16	0.16	1.20	1.10
PPM North	181731	6.2	1931.5	1930.7	0.09	0.09	0.43	0.42
ACRS I	250052	5.1	1950.0	1950.0	0.08	0.08	0.47	0.46
PPM South	197179	6.1	1961.8	1961.6	0.07	0.07	0.30	0.30
PPM N HPS	31841	7.8	1950.3	1948.0	0.07	0.07	0.24	0.25
IRS I	29163	5.1	1947.7	1944.8	0.08	0.09	0.43	0.44
FK5 Extension	03117	12.4	1946.2	1941.9	0.05	0.06	0.24	0.27
FK5	01535		1954.5	1944.2	0.02	0.02	0.07	0.08

where the positional data are given in seconds of arc (annual) and the proper-motion data in seconds of arc per century (centennial). The data for the SAO represent weighted means calculated from values for the individual source catalogs that were provided by H. Jahreiss (private communication). The mean number of catalog positions contributing to the FK5 Extension is a weighted value calculated from the Bright (992 stars) and Faint (2125 stars) parts, while this number cannot be established for the Basic FK5 because of its complicated history (Schwan 1993).

3. PROCEDURE

To replace the old BSC5 positions and proper motions in the correct order, the source catalogs listed in Table 1 were searched in the order given. However, the data found were written to a separate file consisting only of various cross identifiers and the astrometric data. This technique served two purposes, the most important being that each newly created update file was completely independent of the original catalog and thus could be checked easily without touching the full BSC5 itself. Also, each newly created file could be sorted and rearranged, if necessary, before being used as input for the next source catalog search. A source code was created in the first step (the SAO search) and replaced in each succeeding step when astrometric data from a lower ranking source were replaced. Statistics were kept along the way so that an estimate of how the catalog sources had changed could be made. For example, of the

9096 stars in the initial catalog, 9012 with the SAO source had been replaced by PPM data following the first PPM search and only 72 SAO source stars were left. (Twelve stars had position and proper-motion sources other than the SAO because they are not included in the SAO.) After the ACRS search the number of PPM stars had decreased to 7581, while the high-precision PPM subset was 1087 stars. Finally, following the FK5 search, 4432 PPM sources remained and only 67 SAO sources were left, all of which are either close double stars or red variables not included in any of the original list of astrometric source catalogs. Table 2 gives a summary of source codes following the final FK5 run, but before any detailed analyses of sources for double stars, etc.

TABLE 2

Summary of Source Codes for BSC Stars

Source Catalog	Code	Count
SAO	S	0067
PPM	P	4432
ACRS I	A	1032
PPM-H	H	0792
IRS	I	0411
FK5 + Extension	F	2350
Other	O	0120

4. DOUBLE STARS

As in so many other areas of observational astronomy, close double and multiple systems with approximately equally bright components are causing problems. Although it is desirable to include individual positions and proper motions for all stars separated by more than a few seconds of arc, most astrometric catalogs either omit them or give mean positions (center of light) only. Thus, we must either compute individual positions from the center of light (mean) position based on component brightnesses or find meridian circle observations where the components were resolved and measured separately. Certain FK5 systems for which the main FK5 reports center of gravity positions, but where the FK5 appendix gives ephemerides for reduction to the components, were treated easily. Other systems whose orbits are not known or inadequate information is available will not be so easy and mean or less precise astrometric data may be reported. We are currently reviewing these systems one by one and each will be treated individually.

5. EXAMPLES

The following lines illustrate a few entries in the new catalog. Only the astrometric data and their sources, denoted by "C", are shown. Note that the precision of the positions can vary according to the source code.

TABLE 3

Examples of Some Entries in the New Catalogue

HR	Name	α	(J2000.0) δ	μ_α (J2000.0) μ_δ		C
15	21 α And	00 08 23.265	+29 05 25.58	+0.0104	-0.163	F
18		00 08 33.528	-17 34 41.28	+0.0015	-0.044	I
112		00 30 55.117	+77 01 09.47	+0.1016	-0.043	H
291	69 σ Psc	01 02 49.101	+31 48 15.12	+0.0013	-0.035	A
3223	ϵ Vol	08 07 55.833	-68 37 00.07	-0.0052	+0.049	P
3310	22 ϕ^2 Cnc	08 26 46.783	+26 56 03.67	-0.0011	+0.002	S
5459	α^1 Cen	14 39 36.471	-60 50 02.72	-0.4983	+0.699	F
5460	α^2 Cen	14 39 35.166	-60 50 13.23	-0.4983	+0.699	F
5505	36 ϵ Boo	14 44 59.2	+27 04 30.	+0.0092	+0.021	O

REFERENCES

Bastian, U. and Roser, S. (in collaboration with Yagudin, L. I. and Nesterov, V. V.) 1993 Catalogue of Positions and Proper Motions, South, Astronomisches Rechen-Institut, Heidelberg

Corbin, T. E. 1991 International Reference Stars (IRS), U. S. Naval Observatory, Washington, DC

Corbin, T. E. and Urban, S. E. 1990 in IAU Symposium No. 141, Inertial Coordinate System on the Sky, J. H. Lieske and V. K. Abalakin, eds., Reidel, Dordrecht, p. 433

Corbin, T. E. and Urban, S. E. 1990 Astrographic Catalog Reference Stars (ACRS) catalog U. S. Naval Observatory, Washington

Fricke, W., Schwan, H. and Lederle, T. (in collaboration with Bastian, U., Bien, R., Burkhardt, G., du Mont, B., Hering, R., Jahrling, R., Jahreiss, H., Roser, S., Schwerdtfeger, H. M. and Walter, H. G.) 1988 Fifth Fundamental Catalogue (FK5), Part I. The Basic Fundamental Stars, Astronomisches Rechen-Institut, Heidelberg, No. 32

Fricke, W., Schwan, H. and Corbin, T. E. (in collaboration with Bastian, U., Bien, R., Cole, C., Jackson, R., Jahrling, R., Jahreiss, H., Lederle, T. and Roser, S.) 1991 Fifth Fundamental Catalogue (FK5), Part II. The FK5 Extension, New Fundamental Stars, Astronomisches Rechen-Institut, Heidelberg, No. 33

Hoffleit, D. 1982 The Bright Star Catalogue, 4th revised edition, Yale University Observatory, New Haven

Hoffleit, D. and Warren Jr., W. H. 1992 The Bright Star Catalogue, 5th revised preliminary edition, Selected Astronomical Catalogs, Vol. 1, CDROM ADC, NASA/GSFC

Roser, S. and Bastian, U. 1989 in Star Catalogues: A Centennial Tribute to A. N. Vyssotsky, A. G. D. Philip and A. R. Upgren, eds., L. Davis Press, Schenectady, p. 31

Roser, S. and Bastian, U. 1989 Catalogue of Positions and Proper Motions North, Astronomisches Rechen-Institut, Heidelberg

Schwan, H. 1993, private communication

Smithsonian Astrophysical Observatory Staff 1966, Star Catalog: Positions and Proper Motions of 258,997 Stars for the Epoch and Equinox of 1950.0, Publ. of the Smithsonian

Institution of Washington, DC No. 4652, Smithsonian Institution, Washington

DISCUSSION

PHILIP: Is this a project just for you and Dorrit, or have you some help in entering all these new data?

WARREN: The astrometric data are being incorporated into the catalog mostly by computer, so much of the work has already been done by our colleagues who have compiled the new astrometric catalogs in machine-readable form. The only manual labor involved includes the analysis of data for close double and multiple stars, and this work is being done by just the two of us at present.

COHEN: Do you plan to update the BSC by providing MK spectral types for those stars whose types go back to the HD?

WARREN: Many new and revised MK types have already been entered into the catalog (93 percent of the stars have MK types now), but the literature search is still far from complete. Special searches will be made for those stars still lacking any type in the MK system. It is possible that we may ask our classification colleagues to consider working on some of the remaining stars.

RATNATUNGA: At this conference we saw presented the final version of the Yale Parallax Catalog and a preliminary version of the Nearby Stars Catalog. Can you fix a date when the BSC5 (a preliminary version of which I analyzed in 1986) will be finally published?

WARREN: I was afraid that someone would ask that question. Although we are making the current version available in printed (computer output) form upon request, there is still much to do before we are willing to make a final version available in machine-readable and book form. Major tasks include the incorporation of information from the eighth catalog of spectroscopic binaries, the latest photometric catalogs, and a comprehensive literature search since the fourth edition was published. Thus, it will likely be a year or two before a final version can be made available, but that depends strongly on the time available for the BSC5 work, which cannot be foreseen at present because my job situation is not entirely stable.

GALACTIC STRUCTURE TOWARD THE LOW-LATITUDE CLUSTER NGC 3680

R. A. Méndez, V. Kozhrina-Platais, T. M. Girard and W. F. van Altena

Yale University Observatory

ABSTRACT: We present the first comparisons between a newly developed Galactic Structure and Kinematics Model code (GSKM) with magnitude and color counts as well as proper motions in the direction of the open cluster NGC 3680 (b = +17°, l = 287°). We show the adequacy of the GSKM code to reproduce the observations towards this low-latitude field in the anti-galactic rotation direction. Model galactic structure parameters are derived using GSKM by comparing the model preditions with the observed data, after statistical subtraction of the cluster. GSKM is coded in a modular way so that it can easily be modified to extract relevant galactic structure information from current and foreseen databases.

1. INTRODUCTION

Star count studies alone seem unable to characterize definitively the properties of the various components of the Galaxy (e.g., Gilmore et al. 1985, Bahcall and Ratnatunga 1985). Kinematical data are required to compare the predictions of galactic structure models with the observed trends in a multi-dimensional observational space (Gilmore et al. 1989, Bienaymé et al. 1992).

Motivated by several ongoing astrometric projects at Yale we have decided to develop the GSKM code that is to be calibrated self-consistently in a large magnitude range and towards many lines-of-sight through a comparison of our data to the model predictions. In what follows, we present the first comparisons between such a model with magnitude and color counts as well as proper motions in the direction of the low-latitude cluster NGC 3680.

2. SAMPLE CHARACTERISTICS

The plate material consists of a collection of twelve blue plates taken with the 26-inch Yale refractor at Johannesburg and Mt. Stromlo observatories, three blue and three visual plates taken with the CTIO 1-m telescope, and two visual plates taken with the Yale 51-cm astrograph at El Leoncito. All the plates were measured with the Yale PDS laser interferometer/microdensitometer measuring machine. The plates used to compute proper motions span an epoch difference of 45 years.

The master list consists of 3,702 stars found by a digital survey/centering algorithm on a raw scan of the deepest intermediate-epoch 26-inch plate. Individual plates were then scanned in a fine raster mode, based on the master list. Not all the stars on the master list centered in the final centering step using the Yale digital centering algorithm. This leads to having 3,502 stars with V photometry, 3,404 stars with both B and V photometry and 2,698 objects with calculated proper motions. Only the 26-inch plates were used to determine the proper motions.

Workshop on Databases for Galactic Structure
A. G. Davis Philip, Bernard Hauck and Arthur
R. Upgren, eds. p. 223 - 226 © 1993 L. Davis Press

R. A. Méndez
Yale University Observatory
New Haven, Connecticut
06511

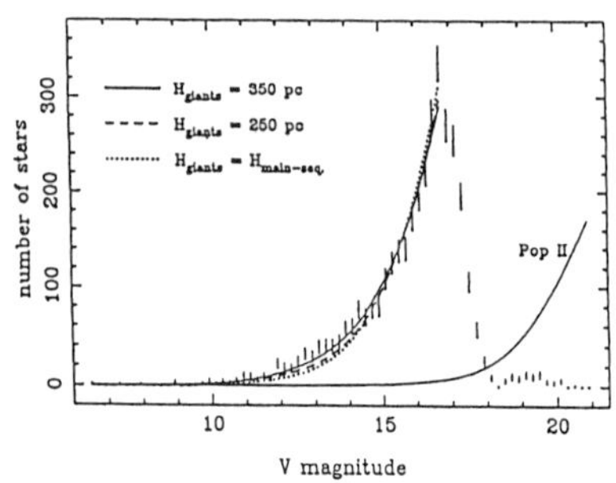

Fig. 1a. Two-Component Model: Magnitude Counts.

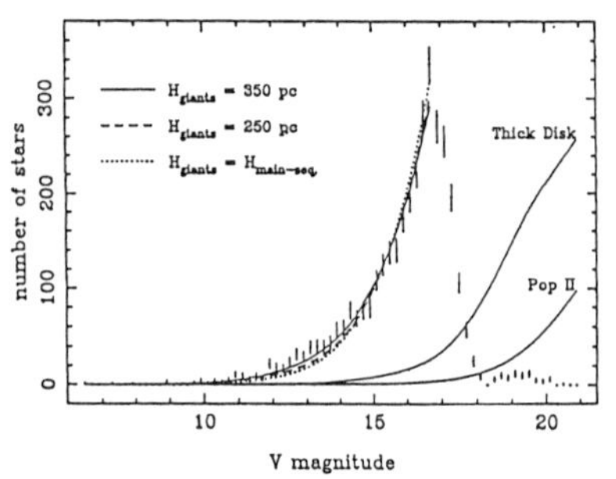

Fig. 1b. Three-Component Model: Magnitude Counts.

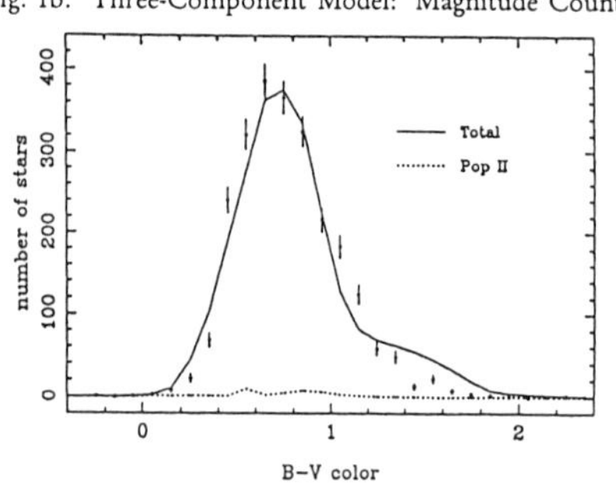

Fig. 2a. Two-Component Model: Color Counts.

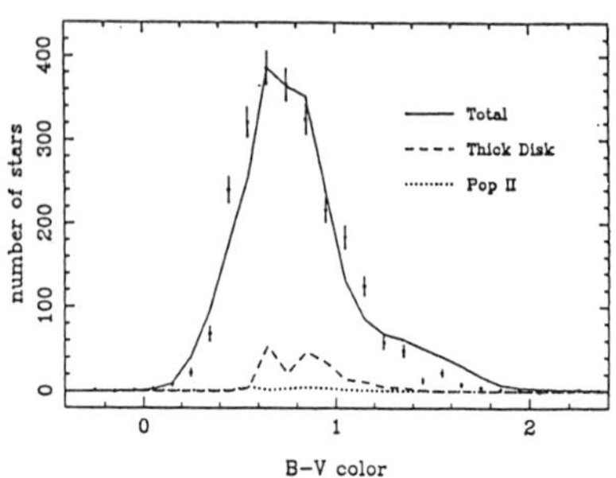

Fig. 2b. Three-Component Model: Color Counts.

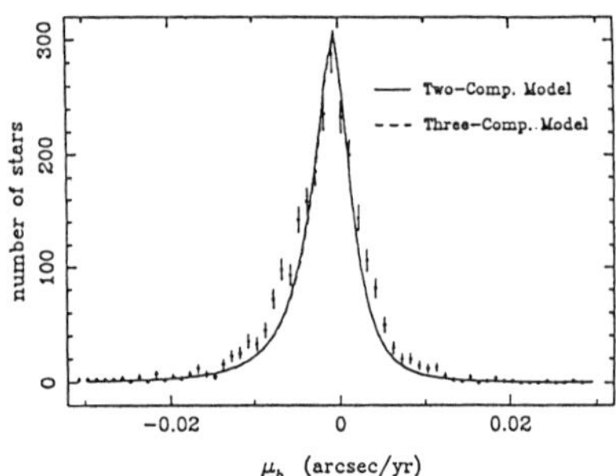

Fig. 3a. μ_b Distribution.

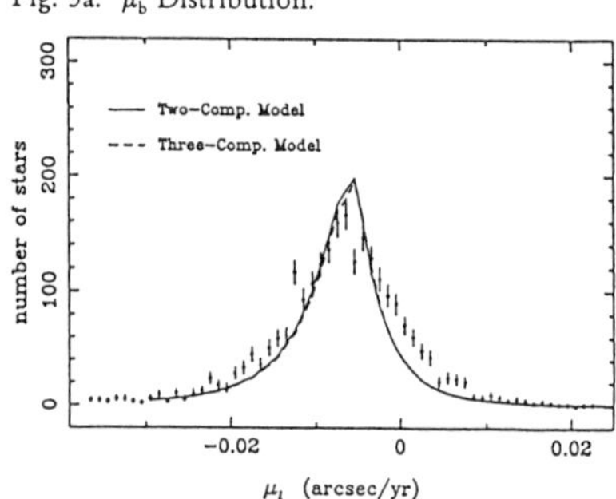

Fig. 3b. μ_l Distribution.

The formal photometric errors for the whole sample are 0.075 in V and 0.098 in B, although they depend strongly on magnitude itself, becoming 0.1 mag for V = 17.0. The formal proper motion error in each coordinate also depends on magnitude, but it is typically around 0.5 mas/yr for magnitudes brighter than V = 17.0.

3. ANALYSIS OF THE SAMPLE

Membership probabilities were computed for all objects in the sample with calculated proper motions, and a statistical removal of cluster members was performed in all of the histograms. Since the cluster is very sparse, the removal amounts only to about 130 objects.

The sample exhibits a sharp cut in the star counts past V = 17.0 (Figs. 1a and 1b). The model predictions will be compared to the data in the $V \leq 16.8$ range where it will be assumed that the sample is complete.

Fig. 1a compares the model predictions using GSKM with the actual star counts for a two-component model "a la Bahcall" (Population I + Population II, Bahcall 1986), while Fig. 1b presents the results for a three-component model "a la Gilmore" (Population I + Thick Disk + Population II, Gilmore et al. 1989).

A reddening E(B-V) of 0.11 mag, at infinity, has been used in the model computations. The reddening is modeled as $E(B-V)(d) = E(B-V)_{\infty}*(1-\exp[|z|/H_{gas}])$, where $E(B-V)_{Inf}$ is the reddening at infinity from Burstein and Heiles (1982) reddening maps, z is the distance from the galactic plane for a point at distance d from the Sun, and H_{gas} is the gas scale-height (110 pc).

We found that our star-counts are rather sensitive to the scale height of giants (H_{giants}). The best fit to the star-counts is obtained by a scale height of 350 pc, while the predicted counts using a scale height of 250 pc or a scale height that follows that of main-sequence stars ($H_{main-seq}$) leads to an underestimation of the number of stars in the range 12 to 14.5 mag.

The low latitude of the field, along with the bright completeness limit, prevents a high contribution from either Thick-Disk or Population II stars to the overall star-counts (Figures 1a and 1b). The two-component model predicts 40 Population II stars with $V \leq 16.8$, while the three component model predicts 20 Population II stars and 190 Thick-Disk stars with $V \leq 16.8$. These numbers are to be compared to the 2,381 observed field stars within the same magnitude limit.

The color counts are presented in Figs. 2a and 2b, for the two- and three-component models respectively. The value that best matches the observations is $E(B-V)_{\infty} = 0.11$. The single-peaked nature of the histograms makes no model substantially different from the other, with the three-component model showing a slightly better fit.

The velocity ellipsoid for this field is oriented with the μ_l component parallel to the galactocentric radius vector (U), while the μ_b component is almost parallel to the component perpendicular to the galactic plane (W). Therefore, the two velocity dispersions Σ_U, Σ_W are separable. Figs. 3a and 3b show the results of the two- and three-component modeling for the sample with $V \leq 16.8$. The velocity dispersions adopted in these model computations were somewhat larger than those for the solar neighborhood (e.g. the adopted velocity dispersions for stars of type G5 V and later were $[\Sigma_U, \Sigma_V, \Sigma_W] = [32, 25, 20]$ km/sec, while the mean values usually quoted are [30, 20, 15] km/sec, respectively (Mihalas and Binney 1981, Ratnatunga et al. 1989).

REFERENCES

Bahcall, J. N. 1986 A.R.A.A. 24, 577
Bahcall, J. N. and Ratnatunga, K. U. 1985 MNRAS 213, 39
Bienaymé, O., Mohan, V., Crézé, Considère, S., Robin, A. C. 1992 A&A 253, 389
Burstein, D. and Heiles, C. 1982 AJ 87, 1165
Gilmore, G., Reid, N. and Hewett, P. 1985 MNRAS 213, 257
Gilmore, G., Wyse, R. G. F. and Kuijken, K. 1989 A.R.A.A. 27, 555
Mihalas, D. and Binney, J. 1981 Galactic Astronomy, Freeman and Co., New York, p. 423
Ratnatunga, K. U., Bahcall, J. N. and Casertano, S. 1989 ApJ 339, 106

MAXIMUM LIKELIHOOD AND THE PROPERTIES OF OPEN CLUSTERS

Stefano Casertano

University of Illinois

Mario G. Lattanzi

Space Telescope Science Institute

and Kavan U. Ratnatunga

Johns Hopkins University

ABSTRACT: A preliminary application of the method of maximum likelihood to the study of open clusters yields a moving-cluster distance estimate for the Praesepe cluster of 140 ±10 pc, significantly more precise than previous distance estimates based on either an assumed velocity or on spectroscopic parallaxes. The space velocity and convergent point are also determined, based exclusively on the kinematics of cluster stars. No *a priori* membership assignment is required for this method.

1. INTRODUCTION

The method of maximum likelihood, described in greater detail in Casertano and Ratnatunga (1993), finds a natural application in the study of nearby open clusters. In this specific case, the method consists of describing the properties of the parent population for both cluster and field stars *separately* as a function of a number of parameters, and maximizing the *combined* probability for all observed stars to belong to such populations. The number of parameters that can be optimized depends on the number of stars included in the sample and on the amount and quality of information available for each of them. When the method is applied to stars within a small region near the center of a cluster, the sample of field stars is usually insufficient to constrain the properties of the galactic stellar population from which they are drawn, and thus it is desirable to obtain the properties of the field stars from other samples, more numerous and better suited to the study of galactic structure. Cluster parameters can then be optimized with minimal loss of information.

In this initial application, we have concentrated our efforts on the *kinematic* properties of the Praesepe cluster, based primarily on proper motions and line-of-sight velocities, to obtain its three-dimensional space velocity, and consequently its distance, independent of any assumptions on the color-luminosity distribution of its stars. For this reason, and to perform a clear-cut test of the potential of the method with respect to cluster kinematics, we have not used color information at this stage. In future applications, we will generally make use of all the information available, although it is occasionally useful to exclude some observables in

Workshop on Databases for Galactic Structure
A. G. Davis Philip, Bernard Hauck and Arthur
R. Upgren, eds. p. 227 - 231 © 1993 L. Davis Press

Stefano Casertano
Physics and Astronomy Dept.
505 Bloomberg Bldg
The Johns Hopkins University
Baltimore, Maryland
21218

order to understand which measured quantities constrain which cluster properties.

2. CONVERGENT POINT: THE CLASSICAL METHOD

The classical approach to find the convergent point for a moving cluster from the proper motions of cluster stars starts from the definition of a condition equation, which translates the ideal requirement that all proper motion vectors for cluster stars point towards the convergent point itself. The optimum convergent point is the point that minimizes the deviations from this ideal condition in some least-squares sense.

The desired condition equation can be written in many different ways, all of which are in principle equivalent *in the absence of errors*. However, different equations lead to different minimizations and thus different least-squares solutions. For example, the condition used by Rasmusen (1921) expresses the requirement that all proper motions point towards the convergent point. In the least-squares sense, this means minimizing the *angles* between the direction to the convergent point and each proper motion. We call this an *angle* method.

On the other hand, the equation used by Green (1985) expresses the requirement that the convergent point must lie on the great circle defined by position and proper motion of each star. In the least-squares sense, this corresponds to minimizing the distance between the convergent point and each great circle; this we call a *distance* method. While both angle and distance methods work well if the errors are sufficiently small, the latter is less satisfactory with large errors. If, for example, the true convergent point is several tens of degrees away from the cluster center, even a relatively small error in angle translates into a large distance between convergent point and great circle, so that, with the distance method, a closer convergent point may erroneously be favored. In fact, the distance method converges properly only if the fractional proper motion errors $\delta\mu/\mu$ are much smaller than the ratio between angular size of the cluster θ and angular distance λ between cluster center and convergent point. With the distance method, the condition $\delta\mu/\mu << \theta/\lambda$ must be satisfied regardless of the number of stars, while the angle method, with a sufficiently large number of stars, will converge, in the absence of systematic errors, to the correct convergent point.

3. MAXIMUM LIKELIHOOD CALCULATION

Even the angle solution is far from optimal. Perhaps the most obvious drawback is the need to assign cluster membership beforehand. Even if membership probabilities are used, these are often based on a preconceived notion of the properties of the cluster, which causes a degree of circularity in the results. Some of the more modern methods used to assign membership probabilities reduce this problem by modeling the proper motion distribution as a combination of Gaussian distributions. Although these methods go a long way towards avoiding selection biases in the determination of cluster properties, they cannot easily be integrated in a self-consistent determination of the convergent point of the cluster, since the very existence of a convergent point implies that the distribution of proper motions for cluster stars is not Gaussian.

Another drawback is that only part of the available information is used. For example, the *direction* of the proper motion is used, but not its *magnitude*, which also carries some information, since the proper motions of stars closer to the convergent point will be on average smaller than those of stars farther away from the convergent point. This effect can be as large as the convergence of the stellar proper motions, although it is blurred to some extent by velocity and distance differences between stars.

Finally, the methods described so far are limited in their application, since they cannot easily account for other cluster properties, such as rotation, density distribution, color-luminosity distribution, et cetera.

Using the maximum likelihood method overcomes these drawbacks. With this method, both the field population of galactic stars and the population of cluster stars are described with a number of parameters, which in general include luminosity function, photometry, kinematics, density distribution, and possibly metallicity. More than one component can be used for each of these populations. The probability of the model (its likelihood) is defined by the probability of drawing such stars as are actually observed from these populations. The better the distribution of stars matches the model, the higher the likelihood of the model. In principle, each and every model parameter can be varied to find the maximum of the likelihood function, which corresponds to the best model given the data. In practice, there may be insufficient information on some of the parameters, because of the size of the sample or because it is restricted to a small region of the sky or is otherwise selected. Therefore, it becomes necessary to predefine the values of some parameters, especially those describing field stars, which are usually underrepresented in samples from near the cluster.

It is however important that all stars in a well-defined region of observable space be included in the sample; if preselection is made to improve the probability of including cluster stars, as may be necessary to optimize observing strategies, this preselection must be made according to well-defined selection criteria, which can then be included in the likelihood calculation.

One of the most obvious advantages of this method is that it does *not* require a priori membership assignments, not even in the form of probabilities; such assignments will in fact bias the resulting group parameters in favor of preselected answers, corresponding to the selection criteria. For each star, the probability of its existence will be the combination of the probabilities for cluster and for field stars; of course, if the star's observable quantities match closely that of cluster stars, then the cluster will contribute the largest probability, while if some quantity does not match the parameters currently chosen for the cluster, its contribution will be less important. The key point is that this assessment is done "dynamically": as cluster properties are varied during the maximization of the likelihood function, so will vary the fractional probability for each star to belong to the cluster or field population.

4. FIRST APPLICATION: KINEMATICS AND DISTANCE OF PRAESEPE

For this first application we have chosen the Praesepe cluster, for which a large amount of information is available. This cluster is just beyond the range of applicability of the classical moving cluster method, and therefore it makes an ideal candidate for the application of our approach. To keep the issues as simple as possible, we have not used in this calculation any of the photometric information which would be used in a full maximum-likelihood calculation.

Instead we have concentrated on the cluster kinematics, as defined by the proper motion and line-of-sight velocity of cluster and field stars in the vicinity of the nominal cluster center. The data used so far are the proper motions and positions from the AGK3U catalog (Bucciarelli et al. 1992) and the positions, proper motions and line-of-sight velocities from the INCA catalog (Turon et al. 1992). Other data have not been incorporated in the solution because of the existence of zero-point differences which have not been included yet in the calculations (see below). We have included all catalog stars within $3°$ of the nominal cluster center $l = 206°$, b $= 32°$ (Strassl 1965), for a total of 348 stars: 271 from the AGK3U and 77 from the INCA. Line-of-sight velocity is available for 27 INCA stars.

For this calculation, no optimization has been attempted on the properties of the field stars, parameters for which have been taken from other work (Ratnatunga and Upgren 1993). As for cluster parameters, we have optimized distance and position of the center, central overdensity and space velocity. The likelihood function is insensitive to the internal velocity dispersion of the cluster, presumably because of the relatively large velocity errors. Thus, the internal velocity dispersion cannot be determined in this preliminary calculation.

The main result of this optimization is what, to the best of our knowledge, is the first kinematic determination of the cluster distance, at 140 ± 10 pc. The components of the velocity are 31.7, 4.2, -29.8 km/s in the line-of-sight, l and b directions respectively, with an error of ~0.5 km/s for the line-of-sight component, and of about 10% in the other components, mostly due to the distance error. The convergent point is at $(l, b) = (208° \pm 2°, -10° \pm 5°)$. The cluster center is at $(l, b) = (205°6, 32°7)$. The central density depends on the assumed density law, which is not well-constrained; thus no precise value can be given based on our calculations. Several tests have been performed to check the validity of our results; among these, we have repeated the minimization while keeping one or more parameters fixed at values different from the optimal ones. Aside from the expected interdependence between distance and tangential components of the velocity, no other abnormality has been found.

Previously reported distances for the Praesepe cluster are 137 pc (Wassink 1927) and 159 pc (in Allen 1973). The former depends on the assumption that the space velocity of the Praesepe cluster is as close as possible to that of the Hyades, while the source for the value in Allen is unclear. No errors are given for either distance estimate. Convergent point positions given in the literature are $(206°, +7°)$ (Rasmuson 1921) and $(207°, -2°)$ (Wassink 1927); both of these are calculated from the distance estimate, rather than derived directly from a study of the proper motions of the cluster.

5. WHAT NEXT?

The results for Praesepe can be improved significantly in two ways. First, more data can be included in the analysis; extensive catalogs of proper motions in the Praesepe region exist in the literature, among which especially interesting are those by Wassink (1927), Russell (1975), and Arthyukhina (1966, 1971); for a more complete list see van Leeuwen (1985). However, there exist very significant zero-point differences between the proper motions from different sources. The problem of systematic differences between catalogs will be addressed self-consistently within our maximum likelihood approach.

Another interesting development is suggested by the possible existence of cluster members at large angular distances, up to 5°, from the cluster center (Mermilliod et al. 1990). Even with relatively less accurate proper motions, such stars can bring enormous leverage to the determination of the distance to the cluster. We would need to enlarge the area around the cluster included in our solution and look for the existance of significant number of such stars from the resulting cluster membership probabilities. The method will also be applied to other nearby clusters. At present, the interesting distance range is between 100 and 200 pc, which would include, among others, the Pleiades and the Perseus moving group. In the near future, access to deeper and/or more accurate surveys of proper motions will probably extend the useful range by at least a factor of two, and of course Hipparcos proper motions and parallaxes, especially with their freedom from most of the systematic effects that plague ground-based measurements, will help improve the accuracy of distance determinations for the nearer clusters.

ACKNOWLEDGMENT

The authors acknowledge the support of NASA grant NAGW-2945.

REFERENCES

Allen, C. W. 1973 Astrophysical Quantities, Athlone Press, London, p. 278

Arthyukhina, N. M. 1966 Trans. Sternberg Astron. Obs. 34, 181

Arthyukhina, N. M. 1971 Comm. Sternberg Astron. Obs. 172, 3

Bucciarelli, B., Daou, D., Lattanzi, M. G.and Taff, L. C. 1992 AJ 103, 1689 (AGK3U)

Casertano, S. and Ratnatunga, K. U. 1993 in Workshop on Databases for Galactic Structure, A. G. D. Philip, B. Hauck and A. R. Upgren, eds, L. Davis Press, Schenectady, p. 167

Green, R. M. 1985 Spherical Astronomy, Cambridge Univ. Press, Cambridge, p. 345

Mermilliod, J. -C., Weis, E. W., Duquennoy, A. and Mayor, M. 1990 A&A 235, 114

Rasmuson, N. H. 1921 Medd. Lunds Astron. Obs., Sr. II, N. 26

Ratnatunga, K. U. and Upgren, A. R. 1993, submitted

Russell, J. 1975 PhD Thesis, University of Pittsburgh

Strassl, H. 1965 in Landolt-Börnstein VI-I, Springer, Berlin, p. 631

Turon, C., et al. 1992 The Hipparcos Input Catalogue, ESA SP-1136, ESA Publication Division, Noordwijk (INCA)

van Leeuwen, F. 1985 in Dynamics of Star Clusters, J. Goodman and P. Hut, eds., Reidel, Dordrecht, p. 579

Wassink, W. J. K. 1927 Publ. Groningen Obs. N. 41

GALACTIC STRUCTURE DATABASE PROJECTS AT VAN VLECK OBSERVATORY

A. G. Davis Philip and A. R. Upgren

Van Vleck Observatory

1. CCD PHOTOMETRY OF FHB AND BHB STARS

1.1 Introduction

A long-term project has been underway locating and measuring horizontal-branch A-type stars in the galactic halo. Schmidt telescope spectral plates have been taken at the Warner and Swasey Observatory, Tonantzintla Observatory in Mexico and at the Michigan Curtis Schmidt telescope at Cerro Tololo in Chile. On these plates all A-type stars were marked and then placed on the photoelectric photometry program to be measured in the Strömgren Four-Color System. Horizontal-branch A-type stars stand out in the (b-y), c_1 diagram because the c_1 indices are 0.2 mag or more above the positions occupied by stars on the Main Sequence. As a further check in the (b-y), m_1 diagram the HB stars are found scattered in a region perpendicular to the Main-Sequence. The regions under investigation are at high galactic latitudes so the fainter stars, with relatively high z distances from the galactic plane, usually check out as FHB stars. Stars brighter than 7th magnitude most often turn out to be the more luminous members of Pop. I, which also fall in similar parts of the four-color diagrams. Over the years many other types of observations have been made, from spectra which gave radial velocities, photometric scans and ultraviolet observations with IUE. Additional information on these investigations can be found in Philip (1989, 1991)

The first major Schmidt survey for early-type stars was that of Slettebak and Stock (1959) at the NGP. Surveys of the SGP were made by Philip and Sanduleak (1968) and Slettebak and Brundage (1971). Recently two much more extensive surveys have been made. A thin-prism survey has been made of regions near the NGP by Pesch and Sanduleak (1986) and a very large Schmidt survey resulted in "A Catalog of Candidate Field Horizontal-Branch Stars. I." (Beers, Preston and Schectman 1988). Very large numbers of candidate FHB stars have been identified in these surveys but they need follow-up photometry to confirm their classifications.

BHB stars in globular clusters have also been measured so as to gain an understanding of where these stars fall in the four-color diagrams. Before 1985 single channel photometry was the method used but for stars of 15 - 16th magnitude the rms errors of a single observation were ± 0.05 which are too high to find meaningful estimates of astrophysical parameters for these stars. This problem was not resolved until CCD photometers came into use. Now, stars at y magnitudes of 17 can be measured with internal rms errors of ± 0.01 mag.

Workshop on Databases for Galactic Structure
A. G. Davis Philip, Bernard Hauck and Arthur
R. Upgren, eds. p. 233 - 236 © 1993 L. Davis Press

233

A. G. Davis Philip
1125 Oxford Place
Schenectady, New York
12308

1.2 CCD Observations

Four-color CCD frames have been taken of M 4, M 5, M 15, M 22, M 55, M 92, NGC 362 and NGC 1904. Most of these clusters have been observed with the smaller 318 x 510 pixel chips which (with a 0.9-m telescope) give an area on the sky of only 2.7 x 4.3 min of arc so it was possible to measure only a small portion of each cluster. The number of BHB stars ranged from 5 to 22 per frame, which is too small a number to do much in the way of statistics. In August of 1992, at Kitt Peak, I was able to use a 2k x 2k chip with an area on the sky of about 17 min of arc (with a 0.9-m telescope) on a side. This area is large enough to cover the entire area of most globular clusters. In M 92 about 100 BHB stars were measured and, if frames are pushed to the limit, star catalogs of up to 10,000 stars can be created. This great increase in the number of stars to be measured results in changes in all steps of the process. These changes include i) the read-out time of the chips is increased, ii) a greater capacity is needed for storage of observational data, iii) improvements are needed in the data reduction and data analysis programs. Archiving of photometric data is also a larger problem.

The longer readout time is mostly a problem while observing the bright standard stars when many short exposures are taken. The great increase in the amount of data taken during an observing run has been handled by using Exabyte tapes instead of 9-track tapes. At the Dominion Astrophysical Observatory Peter Stetson has been adding improvements to the data reduction programs. ALLSTAR automates many of the procedures in DAOPHOT and handles the reduction of stars in very crowded regions better. When the number of stars per frame was in the hundreds it is possible to consolidate the numbering scheme on three frames in a spread sheet by moving the numbers from one row to another as stars are cross-identified. But when the number of stars goes into the multi-thousands this process would would take far too much time. Stetson has developed another set of programs that carry out the cross-identification of stars on all frames, use measures of the standard stars to determine the reduction coefficients of instrumental colors to the standard colors of the photometric system being used. It is necessary to use a program, DAOGROW, which allows you to measure the intensity of stars through a series of "diaphragms" of increasing radii. Each of the standard stars is measured in each of the filters and then one measures the point-spread function (PSF) stars in the same manner. Using these data the program calculates the transformations to the standard system and applies them to all the stellar indices and prints out tables of magnitudes and colors for all the stars measured in the cluster.

The next problem in the process is the final output. The star catalogs are now so large that astronomers often no longer publish the data tables but just present the CM diagram. The data tables are kept in the author's database but this presents problems for those who are interested in obtaining original data for many clusters. Another problem is that in some cases the observational data on tapes are eventually reused for other observers and so all this material is lost. A few observatories have programs to archive CCD frames. For example there is an archive program at DAO where all CCD frames taken with their telescopes are saved but in the United States at the national observatories data tapes are kept only six months and then are used again.

2. PROPERTIES OF A COMPLETE SAMPLE OF A STARS AND USE OF DATA

The A-stars comprise an ideal group to form a sample which is complete within a large volume of space and for which much information is or soon will be available. It is the intention of those responsible for the observing program for HIPPARCOS, the astrometric satellite of the European Space Agency, to include all stars brighter than the eighth visual magnitude, some 40,000 in all. Parallaxes and proper motions of high precision will be determined for all of these as well as the other stars listed in the HIPPARCOS Input Catalogue (INCA). At the completeness magnitude limit of eight, the A-stars (specifically those of classes A0 V through F2 V) within about 100 parsecs should all be present, since their absolute visual magnitudes are between +0.5 and +3.0. The parallax of a star at this distance can be expected to be $+0\overset{''}{.}010 \pm 0\overset{''}{.}0015$ according to recent results from HIPPARCOS. Whenever the error in parallax is less than 15 percent of the parallax itself, the parallax can be considered to be good for two reasons. First it is within the precision limit at which the Lutz-Kelker correction of about ± 0.3 mag and is thus more precise than most secondary distance methods.

The INCA catalogue lists 23,002 stars with spectral types extending from A0 to F2 and luminosity classes from IV to VI. These are distributed as shown in Fig. 1. It can be seen that the stars brighter than 8.0 appear to be essentially complete. It is about at this magnitude that incompleteness sets in (as planned). This is seen from the fact that each half-magnitude interval brighter than 8.0 has about twice as many stars as the next brighter interval, as expected for a uniform distribution. A total of 9015 stars are brighter than 8.0 and these are the ones of particular interest. They should form a virtually complete sample of A0-F2 stars within about 100 parsecs of the Sun.

The databases should contain information for almost all of these stars. Olson (1983) and others have obtained Strömgren photometry for them and most should have radial velocities as well. These data coupled with the results from HIPPARCOS and the spectral types from the Michigan Spectral Survey would provide information for answers to many questions. These include the relative frequency of metallic and peculiar A stars among the normal ones, the degree to which their motions indicate membership in clusters and moving groups and an improvement in the stellar luminosity function between magnitudes 0 and +3.

REFERENCES

Beers, T. C., Preston, G. W. and Schectman, S. A. ApJ 67, 461
Olson, E. H. 1983 A&AS 54, 55
Pesch, P. and Sanduleak, N. 1986 ApJS 60, 543
Philip. A. G. D. 1989 in The Gravitational Force Perpendicular to the Galactic Plane, A. G. D. Philip and P. K. Lu, eds., L. Davis Press, Schenectady, p. 157
Philip, A. G. D. 1991 in Objective-Prism and Other Surveys, A. G. D. Philip and A. R. Upgren, eds., L. Davis Press, Schenectady, p. 143
Philip, A. G. D. and Sanduleak, N 1968 Bol. Ton. Y Tac. 4, 253
Slettebak, A. and Brundage, R. K. 1971 AJ 76, 338
Slettebak, A. and Stock, J. 1959 Astron. Abh. Hamburg, 5

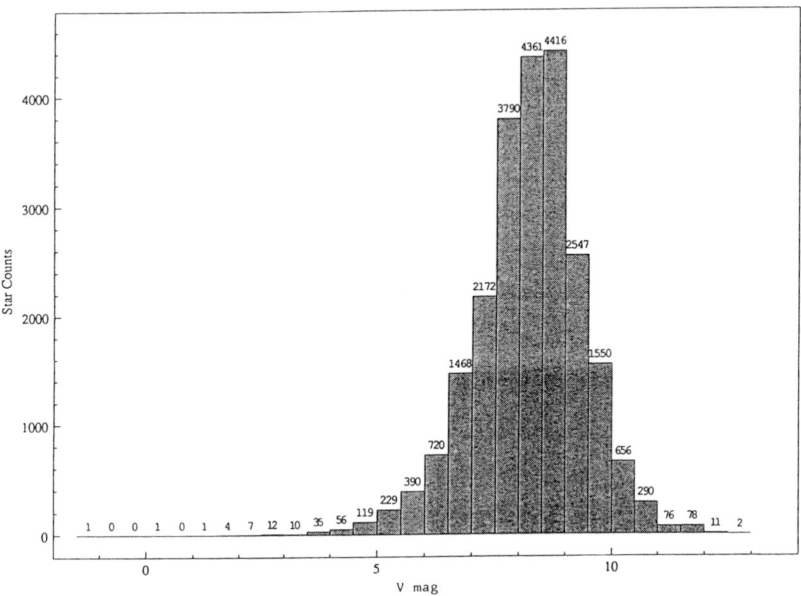

Fig. 1.   The frequency distribution of A0-F2 dwarf stars by apparent magnitude included in the INCA catalogue.

LOCAL VARIATIONS OF THE VELOCITY FIELD FROM NEW STELLAR SAMPLES

E. Oblak

Observatoire de Besançon

M. Moreno

Universitat Politècnica de Catalunya and Universitat de Barcelona

J. Torra

Universitat de Barcelona

ABSTRACT: We analyze the local velocity field trying to state the existence of gradients of the ellipsoid of residual velocities in the galactic plane and in the z-direction. We search these gradients within a 200 pc radius sphere for several groups of stars in the age range $8.2 \leq \log(\text{age}) < 9.5$

1. INTRODUCTION

The observation of stellar motions in the solar neighborhood suggests an ellipsoidal distribution of the stellar residual velocities, giving the shape and orientation of this ellipsoid important clues for our understanding of the dynamics of the Galaxy. The distribution function of the residual stellar velocities may be characterized by its central momenta which, in the traditional approach, have been assumed to be independent of the position. The aim of this work is to look for possible gradients of the velocity ellipsoid. We have developed two statistical methods in order to find up these gradients on the galactic plane and in the z-direction for various age groups of stars. We determine simultaneously for each stellar sample the mean velocity respect to the Sun, the second order moments of the residual velocities and their gradients.

2. THE DATA

The basic sample of stars is selected from the Hipparcos Input Catalogue (Turon et al. 1992). We start gathering main-sequence stars of spectral types B, A and F with radial velocity and proper motions known. Double stars with a separation less than 10" and differences of visual magnitudes less than 4 mag, as well as variable and stars belonging to clusters have been eliminated of our sample. Strömgren photometry is added to the Hipparcos data from the compilation of Hauck and Mermilliod (1990) basically. Perry's catalogue (1991) and our own photometric observations (Figueras et al. 1991) have been also used. Standard errors in radial velocities have been incoporated from the compilation of Barbier-Brossat (1988).

Workshop on Databases for Galactic Structure
A. G. Davis Philip, Bernard Hauck and Arthur
R. Upgren, eds. p. 237 - 240 © 1993 L. Davis Press

237

E. Oblak
Observatoire de Besançon
F-25044 Besançon Cedex
France

238   E. Oblak, M. Moreno and J. Torra

For each star in the sample, effective temperature, gravity, absolute magnitude, and, when possible, metallicity, were derived from uvbyHβ photometry, following basically the procedure described in Figueras et al. (1991). Individual stellar ages have been obtained from the evolutionary models of Schaller et al. (1992) for Pop I stars. Special care has been taken to assign errors for each determined stellar age. The complete method is described in Asiaín (1993). Only stars with relative standard error in age less than 100% have been retained. Photometric distances are considered when trigonometic parallaxes are not known or its relative standard error is greather than 15%. The distances of stars are limited to 200 pc. Finally we limit our working sample to main sequence-stars with log(age) between 8.2 and 9.5. For this preliminary study, six subsamples selected by age have been considered.

3. THE METHODS

3.1 Kinematical Equations

The velocity distribution of stars in the neighborhood of the Sun (that is to say, within a few Kpc of the Sun) can be analyzed by studying radial velocities alone or by combining these with proper motions and known distances. For each star considered, the following quantities and their probable observational errors are available: the radial velocity V_R, the proper motions in galactic longitude $\mu_l \cos b$ and latitude μ_b and the distance r. Proper motion components are derived from proper motions in right ascension μ_α and declination μ_δ, and their errors $\epsilon_{\mu l}$, $\epsilon_{\mu b}$ are obtained by propagation of observational errors $\epsilon_{\mu\alpha}$, $\epsilon_{\mu\delta}$ and ϵ_r taken from the catalogue.

The equations for radial velocity and proper motion expressed in km/s are:

$$V_R = f_R(m, r, l, b) + V_{pec(R)} + \epsilon_{V(R)}$$
$$K_{r\mu l \cos b} = f_l(m, r, l, b) + V_{pec(l)} + \epsilon_{Kr\mu l \cos b}$$
$$K_{r\mu(b)} = f_b(m, r, l, b) + V_{pec(b)} + \epsilon_{Kr\mu(b)}.$$

Radial velocities and proper motions are corrected for galactic rotation. K is a constant equal to 47.41. f_R, f_l and f_b represent the systematic part of the velocity field related to the motion of the Sun (m_{ijk}, $i + j + k = 1$; $m_{100} = U_o$, $m_{010} = V_o$, $m_{001} = W_o$:

$$f_R(m_{ijk}, l, b) = -\alpha_{11} U_o - \alpha_{12} V_o - \alpha_{13} W_o$$
$$f_l(m_{ijk}, l, b) = -\alpha_{21} U_o - \alpha_{22} V_o - \alpha_{23} W_o$$
$$f_b(m_{ijk}, l, b) = -\alpha_{31} U_o - \alpha_{32} V_o - \alpha_{33} W_o$$

α_{ijk} are the elements of the rotation matrix from the (V_R, μ_l, μ_b) base to the galactic heliocentric reference frame with U pointing to the galactic center, V in the direction of galactic rotation and W directed towards the north galactic pole.

The peculiar velocities of the stars ($V_{pec(R)}$, $V_{pec(l)}$, $V_{pec(b)}$)and the observational errors ($\epsilon_{V(R)}$, $\epsilon_{\mu(l)}$, $\epsilon_{\mu(b)}$) with standard deviations (σ_R, σ_l, σ_b) and (s_R, s_l, s_b) respectively, constitute the random term of the eqs. (1). Assuming that these two distributions are independent, the random term will have variance:

$$\tau_R^2 = \sigma_R^2 + s_R^2$$
$$\tau_l^2 = \sigma_l^2 + s_l^2$$
$$\tau_b^2 = \sigma_b^2 + s_b^2.$$

The variances of peculiar velocities (σ^2) are expressed in terms of the velocity ellipsoid parameters (μ_{ijk}, $i + j + k = 2$). If now we assume that the residual velocity ellipsoid varies lineraly with the position, a first order linear expansion around the Sun can be made:

$$\mu_{ijk}(r,l,b) = (\mu_{ijk})_U + \xi \partial \mu_{ijk}/\partial \xi + \eta \partial \mu_{ijk}/\partial \eta + z \partial \mu_{ijk} \, \partial \zeta$$

where: ξ = rcosb cosl (galactic center direction), η = rcosb sinl (galactic rotation direction) and z = rsinb (north galactic pole direction). According to several numerical and theoretical developments, we consider the μ_{101} and μ_{011} momenta to be null, implying that the major axes of the ellipsoid points to the galactic plane.

3.2 The Statistical Methods

3.2.1 Least-Squares Method

The previous equations (1) and (2) are solved using least-squares following a standard method based in the LU decomposition (Press et al. 1985). The normal equations arising from the system (1), which allow us to determine the solar motion, were weighted by $1/\tau_R^2$, $1/\tau_l^2$ and $1/\tau_b^2$ respectively. Those arising from system (2), which give the velocity ellipsoid parameters, with $1/\tau_R^4$, $1/\tau_l^4$ and $1/\tau_b^4$, respectively. Following Gómez (1974) the square of the residuals of eqs. (1) are adopted as the variances τ^2, assuming that no correlation exists between the residuals arising from independent kinematic data. An iterative procedure is followed as a result of the adoption of the previous weights, which make the normal equations be non-linear in the unknowns (Moreno and Torra 1989).

3.2.2 Maximum Likelihood and Monte Carlo Methods

An extension of the method developed by one of us (Oblak, 1983) is used to solve , in a different way, the previous equations. It consist in applying the maximum likelihood method combined with the Monte-Carlo method to solve radial velocities and proper motions equations. A random selection of the solutions is generated and progressively refined until the maximum of the likelihood function is reached. We assume normal distribution for: observational errors in radial velocities and proper motions (ϵ), peculiar velocities (V_{pec}) and random members of equations (1), with respective variances: s^2, σ^2 and τ^2. Note that the normal equations obtained from the method of least squares being weighted by $1/\tau^2$ are the same that the equations of condition obtained from the maximum likelihood method if the peculiar velocities are normally distributed.

4. RESULTS

The two methods have been tested and confronted by its application to various synthetic data sets reproducing realistic local stellar samples. The detailed procedure for the simulation is described in Hernández and Monte (1992). We summarize: the heliocentric distance (r), and the galactic longitude and latitude (l,b) are generated following the density function of stars belonging to the thin disk (Bahcall and Soneira, 1980). A Gaussian error proportional to the heliocentric distance is added to r, with a standard deviation of 0.25r following the discussion of Figueras (1986). The final velocity consists of three contributions: a) the mean velocity field (we have adopted for each component a lineal Oort-Lindblad model with a Oort's constant values of A = 15, B = -10; b) the cosmic dispersions σ_U, σ_V, σ_W = (30, 15, 15)K m/s; c) and the Gaussian observational errors for radial velocities σ_{V-R} = 2 Km/s (Fricke, 1968) and proper motions $\sigma_{\mu lcosb}$ = $\sigma_{\mu b}$ = 0."2/cty (Gliese, 1969).

In a first simulated sample of 500 stars velocity ellipsoid gradients are taken to be null. In the second, we have used the significant values obtained by Oblak (1983) for the group two of their sample. In both cases, the characteristics of the simulation are well recovered by the statistical methods applied. The results obtained by application of the two methods to the same sample show a good agreement. Initial solutions for maximum likelihood method are calculated from the heliocentric velocities of the stars using the usual expressions of central moment of a discrete distribution of velocities. Although the values obtained for the expected gradients are small, some of them can be considered as significant and comparable to those founded by Oblak (1983) for a similar sample. A more detailed study of the samples and an interpretation in terms of the local galactic structure, will be presented in a future paper.

REFERENCES

Asiaín, R. 1993, in preparation
Bahcall, J. N. and Soneira, R. M. 1980 ApJS 44, 73
Barbier-Brossat, M. 1991 A&AS 80, 67
Figueras, F. 1986 Ph.D. Thesis, Barcelona
Fricke, W. 1967 AJ 72, 1368
Gliese, W. 1969 Catalogue of Nearby Stars, Veroff. der Astron. Rechen-Institut, vol. 22, Heildelberg
Gómez, A. 1974 Astron. Nachr. 295, 133
Hernández-Pajares, M. and Monte, E. 1992 IAU Colloquium No. 149, The Stellar Population of Galaxies, B. Barbuy and A. Renzini, eds., Kluwer Academic Press, Dordrecht, p. 430
Hauck, B. and Mermilliod, M. 1990 A&AS 86, 107
Moreno, M. and Torra, J. 1989 in Errors, Bias and Uncertainties in Astronomy, C. Jaschek and F. Murtagh, eds., Cambridge University Press, p. 377
Oblak, E. 1983 A&A 123, 238
Perry, C. L. 1991 PASP 103, 494
Press, P., Teukolsky, S.A. and Wetterling, W. T. 1986 in Numerical Recipes, University Press, Cambridge, p. 502
Schaller, G., Schaerer, D., Maynet, G. and Maeder, A. 1992 A&A 96, 269
Turon C. and 53 collaborators 1992 The Hipparcos Input Catalogue, ESA SP-1136}

THE VELOCITY AND SPACE DISTRIBUTION OF FAINT STARS: A PROGRESS REPORT

Wean Shun Tsay

Institute of Astronomy, National Central University

Phillip K. Lu[*]

Center for Galactic Astronomy
Western Connecticut State University

William van Altena[*]

Yale University

ABSTRACT: The objective of this study is to investigate the distribution of stars in luminosity, space density and velocity at a low Galactic latitude. IIIa-J plates have been obtained with the 4-m Mayall telescope at KPNO and positions and magnitudes measured with the Yale PDS for 5142 objects to the 21st magnitude. CCD Photometry in BVRI, GG385 and GG475 have been obtained in eight fields for color and magnitude calibration. The transformation between BVR, GG385 and GG475 and PDS magnitudes is now complete and will be presented.

1. INTRODUCTION

The objective of this investigation is the determination of absolute parallaxes and proper motions with respect to faint galaxies for a low galactic latitude field. Photographic plates with IIIa-J + GG385 emulsion and filter combination were taken by van Altena at the prime focus of the KPNO 4-meter. In this paper we report on the progress of an investigation of a field at R.A. = 7^h 47^m 31^s and Dec. = + 7° 19.7 (1950) (bII = + 16°2) that is centered on the faint binary white dwarf LP543-32/33. Measurements of selected stars in this field were made some time ago and a parallax of the white dwarf binary was published by van Altena and Mora (1977). Preliminary analysis of the space, luminosity and velocity distribution of faint stars was reported by van Altena et al. (1988).

2. ASTROMETRY

The first field containing twenty-two 4-m photographic plates centered on LP543-32/33 has been completely scanned by Tsay on the Yale PDS with the laser interferometer measuring

[*] Visiting Astronomer, Kitt Peak National Observatory, National Optical Astronomical Observatory, which is operated by the Association of Universities for Research in Astronomy, Inc., under contract with the National Science Foundation.

Workshop on Databases for Galactic Structure
A. G. Davis Philip, Bernard Hauck and Arthur
R. Upgren, eds. p. 241 - 244 © 1993 L. Davis Press

Wean Shun Tsay
Institute of Astronomy
National Central Unversity
Taiwan

system (Lee et al. 1986) which provides a relative position accuracy of approximately 0.3 microns and a magnitude accuracy of about 0^m1 for 7000 objects. The astrometric image positions were computed using the Yale Digital Image Centering algorithms as outlined by Lee and van Altena (1983) and modified by Dr. T. Girard. Using the image shape parameters of the model fit to each image, it is possible to distinguish between uncrowed stars, blends of images and galaxies.

A selection of four hundred reference stars uniformly distributed over the plate and in magnitude was then made for the determination of plate constants, parallaxes and proper motions using the Yale Programs described by van Altena et al. (1986). Stars close to the plate limit that have low signal-to-noise were eliminated from further analysis in our final list of 5142 stars and 55 galaxies. Five plates with poor seeing were also deleted in the final analysis. The distribution of parallax errors as a function of distance from the optical axis shows that the majority of stars within 25 arc-min of the optical axis have parallax errors of 5 (mas), where 1 milli-arc-sec = 0.001 arc-sec (van Altena, et al. 1988).

3. PHOTOMETRY

CCD photometry in the BVRI, GG385 and GG475 systems were obtained by Lu with the #1 0.9-meter telescope at KPNO. Our filter/detector combinations were (IIIa-J + GG385 = mpds), (IIIa-F + GG475, to be obtained), (GG385 + liquid CuSO4 = m385 and (GG475 + red cutoff interference filter = m475).

To calibrate the radial photometric effects on the plate, eight CCD fields evenly distributed in radius were selected. The exposure times used ranged from 10 seconds for the bright stars and Landolt standards of the 12th magnitude, to 40 minutes for a limiting magnitude of 21.

The photometric calibration and reductions were carried out by Natasha Greenman as a senior research project at Western Connecticut State University with a DEC-3100 UNIX system using DIGIPHOT and APPHOT tasks within IRAF. The same procedures were used to calibrate the dwarf F and G stars at the South Galactic Pole by Lu et al. (1992). The photometric calibration for the current project was conducted in two-step processes by first transforming the CCD system to Landolt standards and then transforming to the m475 magnitude and m385-m475 color system.

Since each CCD field has multiple exposures for the transformation of BVR standard stars with short exposures, and intermediate and long exposures for faint objects, many stars may have two to five determinations of both their magnitudes and colors. The agreement between the different determinations was in general good. A preliminary transformation between the CCD m385 magnitude and the mpds photographic magnitude, illustrated in Fig. 1, shows excellent agreement for stars brighter than 19th magnitude, with increasing scatter due to photon noise for fainter stars. A color calibration can not now be carried out due to the lack of IIIa-F plates.

In Fig. 2 we show the star counts in m385 in half magnitude bins based on the preliminary calibration shown in Fig. 1. Note that these magnitudes and counts have not yet been corrected

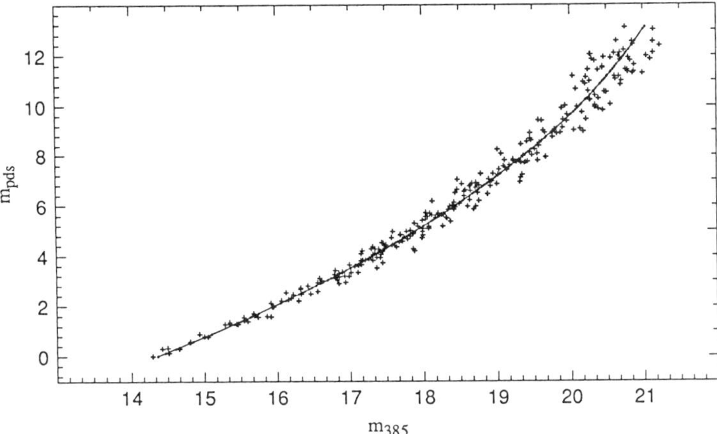

Fig. 1. The relation between our photographic m_{pds} and CCD m_{385} system. See text for the definition of the passbands.

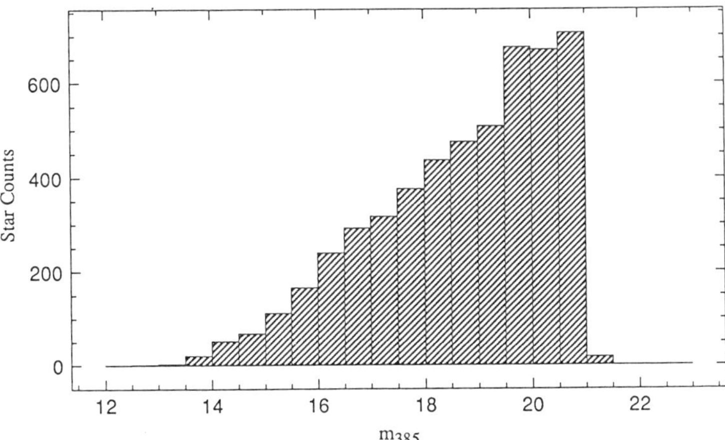

Fig. 2. Star count distribution in m_{385} in half magnitude bins. These counts are based on the preliminary calibration shown in Fig. 1.

for any radial effect in the magnitudes due to the optical field angle distortion.

The final analyses of the astrometric data, luminosity function and space density distribution must wait until we have the IIIa-F photographic plates or large format CCD observations that cover the entire area of a 4-meter photographic plate at the prime focus. A spatial kinematic and metallicity model of the Galaxy, being developed at Yale by R. Méndez (see, for example, Méndez, et al. 1993) will then be used to interpret our observations.

4. ACKNOWLEDGMENTS

We would like to thank Dr. A.A. Hoag and Mr. W.E. Schoening for their encouragement and advice during the photographic observations with the 4-meter at KPNO, Dr. R. Probst for his help in selecting the filter combinations for our CCD observations, Ms. N. Greenman, who carried out the IRAF reduction and analysis, and Dr. D. Guenther for his development of the HORIZON plotting and statistical analysis package which was used to plot the figures in this paper. We would also like to thank the National Science Foundation, for providing funds for the purchase of the PDS at Yale and computer equipment and research support for the investigation at both Yale and Western Connecticut State University.

REFERENCES

Landolt, A. U. 1983 AJ 88, 439
Lee, J -F. and van Altena, W. F. 1983 AJ 88, 1683
Lu, P. K., Miller, J. and Platt, D. 1992 ApJS 83, 203
Méndez, R., Platais, V. and Girard, T. M. 1993 in Workshop on Databases for Galactic
    Structure, A. G. D. Philip, B. Hauck and A. R. Upgren, eds., p. 233
van Altena, W. F., Lee, J. T., Tsay, W. -S., and Lu, P. K. 1988 in IAU Symposium No. 133,
    Mapping the Sky, S. Debarbat, J. A. Eddy, H. K. Eichhorn and A. R. Upgren, eds., Kluwer,
    Academic Pub., Dordrecht, p. 475
van Altena, W. F., Auer, L. H., Mora, C. L. and Vilkki, E. U. 1986 AJ 91, 1451
van Altena, W. F. and Mora, C. C. 1977 BAAS 9, 599

SELECTING NEW CANDIDATE FIELD HORIZONTAL-BRANCH STARS FROM DATABASES[*]

A. G. Davis Philip

Union College and Van Vleck Observatory

Saul J. Adelman

The Citadel

During some cloudy periods on an observing run on the coude' feed at Kitt Peak National Observatory in March (taking high dispersion spectra of some of the brighter FHB stars) we made a database search for additional field horizontal-branch A-type stars by logging in to the Citadel's Vax. First we searched the Hauck Mermilliod (1980) Catalog for all stars with c_1 indices > 1.15 and V magnitudes $\geq$ 7.00 and sent these stars as a list to our university computers. Next the list was sorted by magnitude and then we logged on to SIMBAD in Strasbourg and searched for bibliographic references to all the stars. Six of the stars in the list were already known as FHB stars. 18 stars fell close to the positions occupied by FHB stars in the (b-y), c_1 diagram and these are our candidate stars. Seven stars fell in positions at higher and lower c_1 values. In Fig. 1 the known FHB stars are plotted as (+), the FHB candidates as (x). The Main-Sequence is indicated by the dashed line at the bottom of the figure.

Fig. 2 shows the same group of stars plotted in the (b-y), m_1 diagram. All the stars in this diagram fall well above the dashed line, again indicating the postion of Main Sequence stars. The (b-y), β diagram is shown in Fig. 3. Here all but four of the stars fall below the Main-Sequence relation for A-type stars. In all three diagrams the selected stars fall in the positions occupied by FHB stars and do not follow the Pop I relations.

Of the stars in the figures, twelve are northern hemisphere and twelve are southern hemisphere objects. Thirteen stars have declinations less than 20° from the celestial equator and can be observed from either hemisphere. These stars will be placed on our spectroscopic observing list.

The candidate FHB stars are listed in Table 1. In the table their HD numbers, 1950 positions, V, c_1, m_1 and β indices are given.

---

[*]This research has made use of the Simbad database, operated at CDS, Strasbourg, France.

Workshop on Databases for Galactic Structure
A. G. Davis Philip, Bernard Hauck and Arthur
R. Upgren, eds. p. 245 -248 © 1993 L. Davis Press

A. G. Davis Philip
1125 Oxford Place
Schenectady, New York
12308

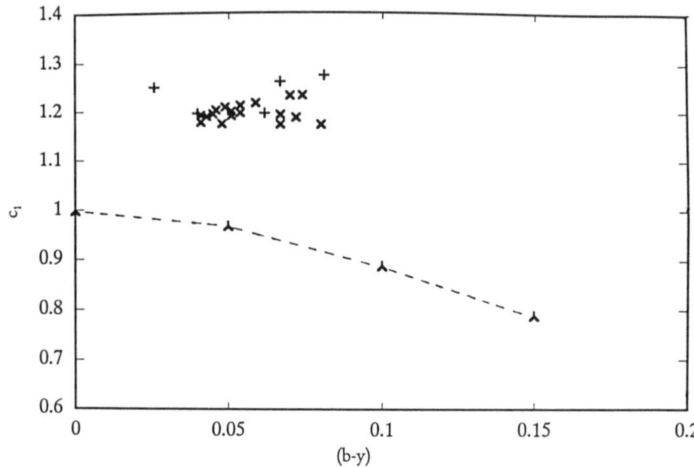

Fig. 1. (b-y) versus c_1 for the stars in Table 1 (X). The dashed line indicates the position of main-sequence Pop I stars. A few known FHB stars which were identified in the search of the database are indicated by (+).

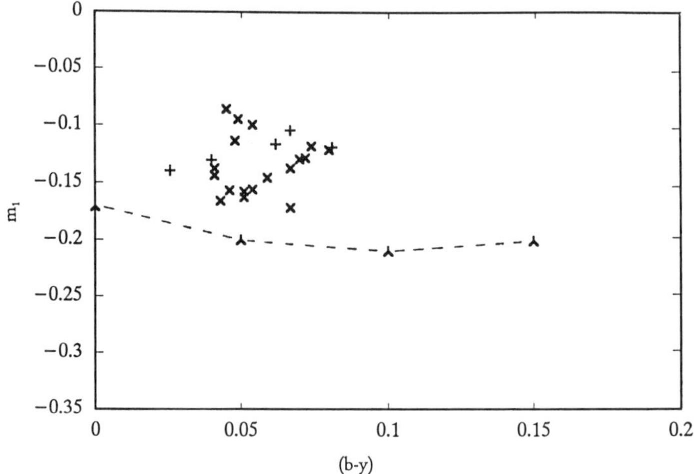

Fig. 2. (b-y) versus m_1 for the stars in Table 1. The same symbols are used as in Fig. 1.

TABLE 1

Candidate FHB Stars

Name	RA	(1950)	Dec	V	(b-y)	c_1	m_1	β
HD 015042	02 23 00.44	+10 16 51.4		07.78	0.080	1.180	0.122	2.835
HD 042999	06 11 38.86	+11 49 46.4		08.80	0.100	1.218	0.100	2.816
HD 047706	06 37 38.02	+08 11 48.5		08.30	0.156	1.203	0.156	2.849
HD 048567	06 41 38.79	+08 22 55.1		08.30	0.049	1.214	0.095	2.769
HD 049224	06 44 55.86	+00 44 55.2		08.28	0.048	1.180	0.114	2.782
HD 053042	07 01 11.90	+05 56 09.8		08.30	0.046	1.208	0.157	2.853
HD 067426	08 05 32.39	+08 22 29.5		07.90	0.059	1.224	0.146	2.820
HD 067452	08 05 23.84	-01 12 12.1		07.66	0.041	1.197	0.138	-
HD 079566	09 12 19.37	-01 22 44.1		07.00	0.051	1.206	0.163	-
HD 083751	09 38 01.08	-00 53 09.6		09.08	0.005	1.186	0.126	-
HD 094509	10 51 25.02	-58 09 25.6		09.08	0.045	1.200	0.086	2.634
HD 100548	11 31 51.85	+48 48 07.0		08.08	0.070	1.240	0.130	2.880
HD 120401	13 47 18.14	-57 44 25.7		08.80	0.074	1.241	0.119	2.914
HD 123664	14 07 00.75	-45 40 41.2		07.63	0.067	1.200	0.138	2.876
HD 128855	14 37 20.75	-40 23 37.8		07.35	0.041	1.182	0.144	2.844
HD 140194	15 40 20.31	-30 22 37.5		07.38	0.067	1.180	0.172	2.914
HD 181119	19 16 04.86	+30 55 48.4		06.70	0.072	1.194	0.129	2.833
HD 185174	19 34 37.66	+31 07 23.2		08.20	0.051	1.197	0.158	2.868
HD 304325	11 26 26.70	-58 07 07.4		08.10	0.064	1.202	0.161	2.867

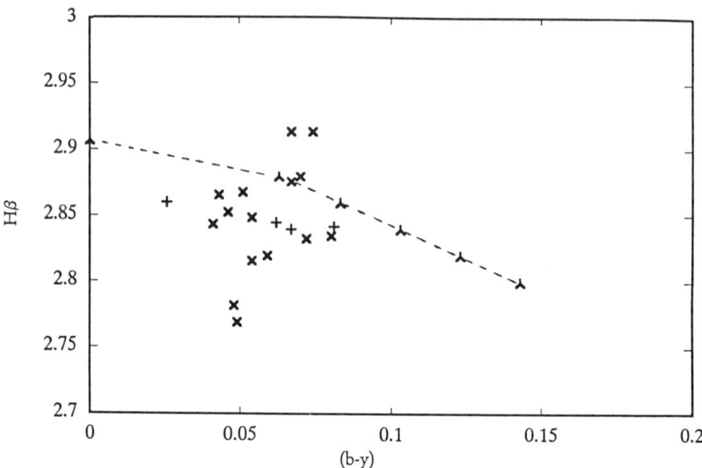

Fig. 3. (b-y) versus β for the stars in Table 1. The same symbols are used as in Fig. 1.

REFERENCE

Hauck, B. and Mermilliod, J. -C. 1980 A&AS 40, 1 (and later electronic versions)

DISCUSSION

PLATAIS: Do you have good positions for your A stars to include them in the SPM input catalog?

PHILIP: Yes, positions for 76 FHB stars selected from the literature were published in VVO Contributions No. 2 in 1984.

WARREN: Has anyone attempted to gather various and sundry lists of horizontal-branch stars into a single compilation?

PHILIP: The paper mentioned in my last answer contains a list of the brighter FHB stars. In recent years, as outlined in the paper, there are some major spectroscopic surveys (Pesch and Sanduleak, and Preston, Schectman and Beers) which are identifying large numbers of possible FHB stars.

LU: I would like to add that there are 3% of the stars out of 3155 F and G star SGP survey are F5 and earlier with (b-y) = 0.24 or bluer. 45 of these stars are in common with the Slettebak and Brundage list. Some of these may turn out to be FHB candidates.

PIER: Lots of BHB candidate stars from the Objective-Prism Survey of Preston, Schectman and Beers have been studied and a sizeable fraction of them are BHB field stars. They published a list of the candidate stars some time ago.

APPENDIX

Astrometric Catalogues: Summary Statistics and Bibliographies
W. F. van Altena
with the collaboration of
K. M. Cudworth, K. Johnston, B. Lasker, I. Platais, and J. L. Russell

Catalogues of Positions and Proper Motions

Ref	Cat.	No. stars	Coverage	T_c	m	σ_p (Tc)	σ_p 1995	σ_μ	Description*
1.	FK5	1,535	All sky	1950	7.0	20	37	0.7	The fundamental reference frame.
2.	FK5 Ext.	3,117	All sky	1944	9.5	55	140	2.6	Faint extension to the FK5.
3.	IRS, I	29,163	All sky	1950	9.5	80	205	4.2	International Reference Stars.
4.	IRS, II	7,064	All sky	1950	9.5	120	315	6.5	Lower precision extension of IRS, I.
5.	ACRS, I	250,052	All sky	1950	10.5	88	230	4.7	Positions and proper motions.
6.	ACRS, II	70,159	All sky	1950	10.5	140[a]	370	7.5[a]	Lower precision extension of ACRS, I.
7.	PPM N	181,731	North	1931	10.5	100	290	4.3	Positions and proper motions, north.
8.	PPM S	197,179	South	1962	10.5	73	120	3.0	Positions and proper motions, south.
9.	CPC2	276,131	South	1967	10.5	54	-	-	Positional catalogue.
10.	HIC	118,209	All sky	var	12.5	-	300	4	Hipparcos Input Catalogue (positional error for 1990).
11.	AGK3	183,173	North	1944	10.5	210	550	10	Positions; motions from AGK3 - AGK2.
12.	AGK3U	170,464	North	1951	10.5	118	280	5.8	Positions; motions from AGK3 & AGK2 + revised Ref. 18.
13.	SAOSC	258,997	All sky	1930	9	200	990	15	Accuracy at publication date, not mean epoch.
14.	FHST	219,859	All sky	var	9.0	var	var	var	HST star tracker catalogue, complete to 8.5.
15.	SKYMAP	248,558	All sky	1930	9	200	990	15	NASA operations catalogue.
16.	CPC	454,875	South	var	10	var	var	var	The Cape Photographic Catalogues for 1950.
17.	Yale Zone	221,338	All sky	var	10	var	var	var	The Yale Zone Catalogues, integrated version available.
18.	HST GSC	2x10e7	All sky	1980	16	250	-	-	HST Guide Star Catalogue, plates from Schm. Surv. 35,42,43.
19.	KSZ	21,817	Dec≥-30	1960	9	150	-	-	Included in the IRS Catalogue (Ref. 3 & 4).
20.	BKAT	4,949	All sky	1969	5.9	80	90	1.7	Bright star catalogue of positions and proper motions.

Explanation of the columns:

*   The column labeled **Ref**. gives a number which may be found in the Bibliography, where the detailed source for each catalogue (**Cat.**) may be found, additional data about the catalogue include, the approximate number of stars (**No. stars**), **Coverage** gives the declination limits, T_c the central epoch, **m** the magnitude limit, $\sigma_p(Tc)$ the error in milli-arc-sec (mas) of a position at the central epoch, $\sigma_p(1995)$ the error of a position in mas at 1995, σ_μ the error of a proper motion in mas/yr, and **Description** is a brief description of the catalogue. Machine readable versions of all catalogues may be obtained at the Astronomical Data Center at Goddard Space Flight Center and at Data Centers in other countries. Acronyms are defined in the Bibliography.

a   Estimated value.

Future Catalogues and Surveys in progress

Ref	Cat.	No. stars	Coverage	T_c	m	σ_p (Tc)	σ_p 1995	σ_μ	Description
21.	Hipparcos	118,209	All sky	1991	9	2	8	2	Some stars to m = 13 with lower accuracy.
22.	TYCHO	1,100,000	All sky	1991	12.0	30	-	-	The Tycho Catalogue from Hipparcos.
23	TAC	1,000,000	All sky	1989	12.0	50	50	3	Ref. 22 plus re-reduced Astrographic Catalogue positions.
24	GAC	4,000,000	All sky	1954	12.5	200	350	7	Ref. 18 plus re-reduced Astrographic Catalogue positions.
25.	IFRS	3,000	All sky	-	13	-	-	-	Intermediate Fund. Ref .Stars (9.5 $\leq V \leq$ 13) -(finding list).
26.	NPM	150,000	+90 to -23	1968	18	120	180	5	Absolute proper motions wrt galaxies, outside Milky Way.
	NPM GP	150,000	Gal. Plane	1968	18	120	180	5	Absolute proper motions wrt galaxies, inside Milky Way.
27.	SPM	1,000,000	-22 to -90	1980	18	100	125	5	Absolute proper motions wrt galaxies.
28.	FON	36,000,000	+90 to -15	1990?	16	150	-	-	Possibly completed by year 2000.
29.	EKAT	1,000,000	+20 to -20	1993?	12	150	-	-	Equatorial catalogue, observations 60% completed.
30.	AC2	4,000,000	All sky	?	13	150	?	?	First epoch is Astrogr. Cat. + new epoch, project on hold.

See Previous Table for a description of the column headings.

Sources for new positions (See Triennial Reports of IAU Commissions 8 and 24 and the Publications of the respective Observatories)

Meridian Circles: Carlsberg Automatic Meridian Circle (-45 to +90, La Palma, Islas Canarias, published annually by Copenhagen Univ. Obs., Royal Greenwich Obs. and Real Inst. y Obs. de la Armada, San Fernando); U.S. Naval Observatory (All sky, Washington and Black Birch, N. Z.); additional northern Meridian Circles in Belgrade, Bordeaux, Pulkova, San Fernando (Spain, transferred to El Leoncito, Argentina in 1994), Shanghai, Yunnan (China) and southern Meridian Circles in San Juan (Argentina) and Santiago (Chile).

Astrolabes: Northern Astrolabes exist in Beijing (currently in San Juan, Argentina), Belgrade, Bucharest, CERGA (France), Paris, Pulkova, San Fernando (Spain), Shanghai, Washington, and Yunnan (China), while Southern Astrolabes exist in Rio Grande and San Juan (Argentina), Santiago (Chile) and Sao Paulo (Brazil).

Major Schmidt Surveys and their Digitization

(Prepared in collaboration with Barry M. Lasker, Space Telescope Science Institute)

Survey	Name	Color	Emulsion	Filter	Dec Centers	Corr.	Cent.	N	Epoch	Archive	Distrib.	Mag.	Ref.
31.	POSS-I	O	103a-O	none	+90, -30	old	6	936	1950-58	CIT	CT GrArt	21	31
32.	POSS-I	E	103a-E	red	+90, -30	old	6	936	1950-58	CIT	CT GrArt	20	31
33.	Luyten	E	103a-E	red	+90, -30	old	6	936	1962-70	Minn	none	21	32,46
34.	Hoessel	IR	IV-N	WR88a	$|b| \geq 10$	old	6	80	1975-79	CIT	none	19	37
35.	Pal Quick V	V	IIa-D	WR12[1]	+90, +6	old[3]	6	613[2]	1982	ST ScI	none	19	33
36.	POSS-II	J	IIIa-J	GG385	+90, 0	new	5	894	1987-	CIT	ESO	22	34
37.	POSS-II	F	IIIa-F	RG610	+90, 0	new	5	894	1987-	CIT	ESO	21	34
38.	POSS-II	N	IV-N	RG9	+90, 0	new	5	894	1987-	CIT	ESO	19	34
39.	USNO QJ	J	IIIa-J	GG385	+90, 0	new	5	894	1987-	USNO	none	13	34,35
40.	ESO	B	IIa-O	GG385	-90, -20		5	606	1973-78	ESO	ESO	20	39
41.	ESO	R	IIIa-F	RG630	-90, -20		5	606	1978-90	ESO	ESO	21	36
42.	SERC-J	J	IIIa-J	GG395	-90, -20	old[3]	5	606	1974-87	ROE	ESO	22	41
43.	SERC-EJ	J	IIIa-J	GG395	-15, 0	old[3]	5	288	1979-	ROE	ROE	22	41
44.	SERC-ER	R	IIIa-F	OG590	-15, 0	old[3]	5	288	1984-	ROE	ROE	21	41
45.	AAO SES-R	R	IIIa-F	OG590	-90, -20	new	5	606	1990-	ROE	none	21	38,49
46.	SERC-I/SR	IR	IV-N	RG715	$|b| < 10$	old[3]	5	163	1978-85	ROE	none	19	42
47.	SERC-I	IR	IV-N	RG715	$|b| > 10$	old[3]	5	731	1980-	ROE	none	19	38

Explanation of the columns:
This and the following table summarize the data available to the authors on the status of the major Schmidt telescope surveys and the process of their digitization as of May 1993. The columns in the first table list sequentially, a reference number for the **Survey** which is used in the following "**Digitization of Schmidt Plates**" table, an acronym (**Name**) for the survey, **Color** passband, **Emulsion** and **Filter** used, inclusive Declination **Centers** for the survey, the Schmidt **Corr**ector used (note that this changes for some of the surveys), the distance in degrees between the declination zones (**Cent.**), **N**umber of plates, **Epoch** of the plates, location of the plate **Archive**, source of **Distrib**ution for copies, if any, **Mag**nitude limit, and the **Ref**erence number in the bibliography where more information may be found.

Notes:
    Field size is 6.5 deg. for Palomar (31 - 39) & UK Schmidts (42 - 47), 5.5 deg. for ESO Schmidt (40 - 41).
1   Most plates taken with the Wratten 12 and a few with a GG495.
2   There are 583 plates in the main survey and 30 in the +3 degree zone, to cover the tilt between the 1855 and 1950 grids.
3   Corrector changed during observational program.

Digitization of Schmidt Plates[4]

Site	Surveys	Pixel	Status	Ref.	Distribution	Net Access
STScI	32, 42, 43	1.7"	done	45	Compressed CD ROM set, 1993-94	no
ROE	40, 41, 42	1.1"	done	40, 47	Compressed CD ROM and at selected sites	no
	36-38, 45	1.1"	scanning	45	Compressed CD ROM and at selected sites	no
Cam	31, 32, 42, 43	0.5"	done	48	Via collaboration	no
Minn	31, 32	0.5"	done	43, 44	Internet	yes
	33, 36, 37, 38	0.5"	scanning	44	Internet	yes
USNO	31,32,36-39	0.9"	scanning	35	Form not determined	yes
STScI	36, 37, 38	1.0"	scanning	45	Compressed CD ROM set, beginning circa 1995	HST DADS

Explanation of the columns:
The above table summarizes the status of Schmidt plate digitization known to the authors as of May 1993. The columns list information for each "Digitization Center" giving an acronym for the center (**Site**) , the **Surveys** being digitized (the number refers to the (**Survey**) previous table), the digitization **Pixel** size in arc seconds, the **Status** of the project (either completed or in the process of being scanned), **References** to the project in the bibliography, plans for **Distribution** of the data, and whether or not the data will be available over electronic networks of some kind (**Net Access**).

Notes:
For more details see: Digitized Optical Sky Surveys Newsletter and IAU Working Group on Wide-field Imaging Newsletters, both edited by H. T. MacGillivray (ROE).

The following three tables summarize data on catalogues providing surveys of relative proper motions, catalogues compiling stars with proper motions larger than some limit, and catalogues that compile varieties of data for a specific type of star, e.g. all bright stars, stars nearer than some distance, etc.. The columns in each table are similar and list respectively, a **Reference** to the bibliography where the catalogue may be found, an acronym of the **Catalogue**, number of stars (**No. stars**), declination **Coverage**, date of publication (**Tpub.**), magnitude limit, proper motion or some other kind of limit (μlim.), and a brief **Description** of the catalogue.

Modern Relative Proper Motion Surveys

Ref	Cat.	No. stars	Coverage	Tpub.	m	μlim.	Description
50.	BPM	94263	South+	1963	15	0.10	Bruce Proper Motion Survey, 1928-63.
51.	Lowell	8991	North	1971	16	0.26	Lowell Proper Motion Survey, 1958-70.
52.	Lowell	2758	South	1978	16	0.20	Lowell Southern section. 1959-78.
53.	LP	450000	North-	1979	21	0.08	Luyten Palomar 48-Inch Schmidt Survey, 1963-77.
54.	Calan	830	South	1989	21	0.15	Selected regions, Maksutov at Cerro El Roble, Chile.

Relative Proper Motion Summary Catalogues

Ref	Cat.	No. stars	Coverage	Tpub.	m	μlim.	Description
55.	NLTT	58855	All sky		All	0.18	Stars with $\mu \geq 0.18$ "/yr.
56.	LHS Cat.	4447	All sky	1979	All	0.50	Stars with $\mu \geq 0.50$ "/yr., 2nd ed.
57.	LHS Atlas	3040	All sky	1979	All	0.50	Stars with $\mu \geq 0.50$ "/yr. Finding Charts for LHS Cat.

Stellar Compilation Catalogues

Ref	Cat.	No. stars	Coverage	Tpub.	m	Limits	Description
60.	BSC V	9110	All sky	1992	6.5	$V \leq 6.5$	The Bright Star Catalogue, 5th ed. Stars with $V \leq 6.5$.
61.	BSC Supp.	2603	All sky	1983	7.1	$V \leq 7.1$	The BSC Supplement. Stars with $V \leq 7.1$ photoelectric.
62.	CNS 3	3803	All sky	1993	All	$\pi \geq 0.040$	Catalogue of Nearby Stars, 3rd prelim. ed. Stars with $\pi \geq 0.040$.
63.	YPC	7883	All sky	1993	All	-	Gen. Cat. Trig. Parallaxes, All stars with measured π.
64.	WDS	73610	All sky	1984	All	-	Washington Double Star Catalogue. All measures.
65.	CHARA	8976	All sky	1988	10	-	Interferometric measures of double stars. 2nd Cat.
66.	Orbits	847	All Sky	1983	All	-	Visual Double Star Orbit Catalogue.

References:

More detailed compilations may be found in:

Collins, M. 1977, "Astronomical Catalogues 1951 - 1975". London INSPEC Series #2.

Eichhorn, H. 1974, "Astronomy of Star Positions", (Frederick Ungar Publ. Co., New York).

Hawkins, G. S. and Rosenthal, S. K. 1967, "5,000 and 10,000-Year Star Catalogs", Smithsonian Contrib. to Astrophysics, 10, 2.

Jaschek, C. 19xx, Standard Values and Information in Data Banks, IAU Symposium #111, p. 331.

Sevarlic, B. M., Teleki, G., and Szadeczky-Kardoss, G. 1978, Bibliography of the Catalogues of Star Positions". Publ. Dep. Astr. Belgrade, No. 7, 69.

Sevarlic, B. M., Teleki, G., and Knezevic, Z. 1982, Bibliography of the Photographic Catalogues of Star Positions, Publ. Obs. Belgrade, 29, 71.

Bibliography

1. FK5 — Fricke, W., Schwan, H., Lederle, T., Bastian, U., Bien, R., Burkhardt, G., du Mont, B., Hering, R., Järling, R., Jahreiß, H., Röser, S., Schwerdtfeger, H.-M., Walter, H., 1988, Fifth Fundamental Catalogue (FK5). Part I). The Basic Fundamental Stars. Veröff. Astron. Rechen-Instituts, Heidelberg, 32 (and Ref. 67).

2. FK5 Ext. — Fricke, W., Schwan, H., Corbin, T., Bastian, U., Cole, C., Jackson, R., Järling, R., Jahreiß, Lederle, T., Röser, S., 1991, The FK5 Extension: New Fundamental Stars, Veröff. Astron. Rechen-Instituts, Heidelberg, 33.

3. IRS, I — Corbin, T. 1991, U. S. Naval Observatory, (and Ref. 67). IRS = International Reference Stars.

4. IRS, II — Corbin, T. 1991, U. S. Naval Observatory, (and Ref. 67). IRS lower precision data.

5. ACRS, I — Corbin, T., and Urban, S. E. 1991, U. S. Naval Observatory, (and Ref. 67). ACRS = Astrographic Catalogue Reference Stars. This catalogue and Ref. 6 are intended to be used for the re-reduction of the Astrographic Catalogue, so they contain no AC positions and emphasize the inclusion of early epoch positions; both hemispheres are averaged together to form the average errors.

6. ACRS, II — Corbin, T., and Urban, S. E. 1991, U. S. Naval Observatory, (and Ref. 67). ACRS lower precision data.

7. PPM N — Röser, S., and Bastian, U., 1991 PPM Star Catalogue, Vols. I & II, Astron. Rechen-Instituts, Heidelberg (and Ref. 67). PPM = Positions and Proper Motions. This catalogue and Ref. 8 are intended to include all available data and hence provide the most precise positions and proper motions; all data are averaged to give the average errors rather than high and low precision data separately as for Refs. 5 & 6.

8. PPM S — Röser, S., and Bastian, U., 1993 PPM Star Catalogue, (The Southern Hemisphere) Astron. Rechen-Instituts, Heidelberg.

9. CPC2 — de Vegt, C., Murray, C. A., Zacharias, N., Nicholson, W., Penston, M. J., and Clube, S. V. M., 1992, A & A Supplements 1992 CPC2 = The Second Cape Photographic Catalogue.

10. HIC — Turon, C., et al., 1992, HIC = The HIPPARCOS Input Catalogue, ESA SP-1136.

11. AGK3 — Dieckvoss, W., Kox, H., Gunther, A,. Brosterhus, E., 1975, The AGK3 Catalogue of Positions and Proper Motions North of -2º.5 Declination, Vols. 1-8, Hamburg-Bergedorf (and Ref. 67). AGK3 = Third Catalogue of the Astronomisches Gesellschaft.

12. AGK3U — Bucciarelli, B., Daou, D., Lattanzi, M. G., and Taff, L. G., 1992, Astron. J. 103, 1689. AGK3U = Update to Ref. 11 using a re-reduction of Ref. 18 for third epoch positions.

13. SAOSC — Smithsonian Institution, 1966, SAOSC = Smithsonian Astrophysical Observatory Star Catalogue, Washington, D.C. (and Ref. 67).

14. FHST — Yang, T.-g., Girard, T., Lee, J. T., and van Altena, W. F. 1992, FHST = A New Fixed Head Star Tracker Catalogue for the HST, in Ref. 68, p. 235.

15. SKYMAP — NASA Goddard Space Flight Center, Greenbelt. An operations catalogue for guidance and tracking NASA spacecraft.

16. CPC — CPC = Cape Photographic Catalogue for 1950, Publications of the Royal Observatory, Cape of Good Hope.

17. Yale Zone — Yale Photographic Catalogues, 1926-83, Trans. Astron. Obs. Yale University, Vols 3-32; and Yale Zone Catalogues - Integrated Version (Ref. 67).

18. HST GSC — Lasker, B. M., Sturch, C. R., McLean, B., Russell, J. L., Jenkner, H., Shara, M. 1990, Astron. J. 99, 2019, and CD-ROM as GSC 1.0; an updated version GSC 1.1 was issued on CD-ROM on 1 August 1992. Positional error is the average *local* precision.

19.	KSZ	Proposed by M. S. Zverov. This incomplete catalogue was incorporated into the IRS (Ref. 3 & 4). KSZ = Faint Star Catalogue.
20.	BKAT	Chrutskaya, E. V. 1985, Astron. J. U. S. S. R. 62, 605. BKAT = Bright Star Catalogue.
21.	Hipparcos	Publications appear in conference proceedings and the ESA SP Series.
22.	Tycho	Publications appear in conference proceedings and the ESA SP Series.
23.	TAC	Roeser, S. 1993, in Ref. 70. TAC = Tycho second epoch positions with Astrographic Catalogue first epoch positions re-reduced using Hipparcos stars (Ref. 21) as the reference frame.
24.	GAC	Urban, S. E. 1993, in Ref. 70. GAC = HST GSC (Ref. 18) second epoch positions with Astrographic Catalogue first epoch positions re-reduced using ACRS stars (Ref. 5 & 6) as the reference frame.
25.	IFRS	Corbin, T. E., and Urban, S. E., 1990, Faint Reference Stars, in Ref. 69, p. 433. See also Report of the Working Group on Star Lists, in Ref. 68, p. 142. Note that this is a finding list for observers and not a catalogue. IFRS = Intermediate Fundamental Reference Stars.
26.	NPM	Klemola, A. R., Jones, B. F., and Hanson, R. B., 1987, Astron. J. 94, 501. See also Ref. 68, p. 235 and Klemola, et al. in Ref. 71. NPM = Northern Proper Motion survey.
27.	SPM	van Altena, W. F., Girard, T., Lopez, C. E., Lopez, J. A., and Molina, E. 1990, The Yale-San Juan Southern Proper Motion Program (SPM), in Ref. 69, p. 419. See also López, C. E. and Platais, I. in Ref. 68, p. 238; Platais, et al. in Ref. 70; and van Altena, et al. in Ref. 71. SPM = Southern Proper Motion survey.
28.	FON	Kolchinsky, I. G., Garilov, I. V., and Onegina, A. B. 1978, in IAU Colloq. 48, Modern Astrometry, edited by F. V. Prochazka and R. H. Tucker, (Univ. Obs. Vienna), p.479. FON = Four-Fold Coverage of Northern Sky.
29.	EKAT	Polojentsev, D. D., Potter, H. I., Yagudin, L. I., and Zalles, R. F. 1991, in Proceedings of the First Spain-USSR Workshop on Positional Astronomy and Celestial Mechanics, edited by A. López Garcia, R. F. López Machi, and A. G. Sokolsky, (Univ. Valencia, Obs. Astron.), p. 63. EKAT = Equatorial Catalogue.
30.	AC2	Proposed by H. I. Potter. This catalogue will use the Astrographic Catalogue as the first epoch and new observations with similar astrographs for the second epoch. The project has been suspended for the time being.
31.	POSS-I	Minkowski, R. L., and Abell, G. O. 1963, in Basic Astronomical Data, ed. K. Aa. Strand; Volume III of Stars and Stellar Systems, ed. Kuiper and Middlehurst, p. 481. POSS = Palomar Observatory Sky Survey.
32.	Luyten	Luyten, W. J., 1963-87, Proper Motion Survey with the Forty-Eight inch Schmidt Telescope, University of Minnesota, Minneapolis.
33.	HST GSC	Lasker, B. M., Sturch, C. R., McLean, B., Russell, J. L., Jenkner, H., Shara, M. 1990, A.J. 99, 2019. HST GSC = Hubble Space Telescope Guide Star Catalogue.
34.	POSS-II	Reid, I. N., Brewer, C., Brucato, R. J., McKinley, W. R., Maury, A., Mendenhall, D., Mould, J. R., Mueller, J., Neugebauer, G., Phinney, J., Sargent, W. L. W., Schombert, J., and Thicksen, R. 1991, Publ. A.S.P. 103, 661.
35.	USN PMM	Monet, D. G., and Westerhout, G. 1989, in Digitized Optical Sky Surveys, ed. C. Jaschek, in Bull. d'Info. du CDS, Number 37, Obs. de Strasbourg, p. 75. The USNO QJ is a two minute exposure on unhypersensitized IIIa-J emulsions to be used for the determination of positions of bright stars. USN PMM = U. S. Naval Observatory Precision Measuring Machine.
36.	ESO-R	West, R. M. 1984, in IAU Colloq. 78, p. 13. ESO-R = European Southern Observatory Red survey.
37.	Hoessel-IR	Hoessel, J. G., Elias, J. H., Wade, R. A., and Huchra, J. P. 1979, Publ. A.S. P. 91,41. Palomar Infra-red survey.

38. AAO-SES   Morgan, D. H., Tritton, S. B., Savage, A., Hartley, M., and Cannon, R. D. 1992, in Digitized Optical Sky Surveys, ed. H. T. MacGillivray and E. B. Thomson (Dordrecht: Kluwer), p. 11. AAO-SES = Anglo Australian Observatory Second Epoch Survey.

39. ESO-B   West, R. M., and Schuster, H.-E. 1982, A&A Suppl. Ser. 49, 577. ESO-B = ESO Blue survey.

40. ROE dig.   Yentis, D. J., Cruddace, R. G., Gursky, H., Stuart, B. V. Wallin, J. F., MacGillivray, H. T., and Collins, C. A. 1992, in Digitized Optical Sky Surveys, ed. H. T. MacGillivray and E. B. Thomson (Dordrecht: Kluwer), p. 67. ROE dig. = Royal Observatory Edinburgh digitizing program.

41. SERC-J/R   Cannon, R. D. 1984, in IAU Colloq. No. 78, "Astronomy with Schmidt Telescopes", edited by M. Capaccioli (Reidel), p. 25. SERC-J/R = Science and Engineering Research Council J and R survey.

42. SERC-I/SR   Hartley, M., and Dawe, J. A. 1981, Proc. Astron. Soc. Austr. 4, 251. SERC-I/SR = Science and Engineering Research Council Infra-red and Short Red survey.

43. Minn APS   Pennington, R. L., Humphreys, R. M., Odewahn, S. C., Zumach, W. A., andThurmes, P. M. 1993, PASP 105, 521; see also Pennington, R. L., Humphreys, R. M., Odewahn, S. C., Zumach, W. A., and Stockwell, E. B. 1992, in Digitized Optical Sky Surveys, ed. H. T. MacGillivray and E. B. Thomson (Dordrecht: Kluwer), p. 77. Minn. APS = University of Minnesota Automated Plate Scanner.

44. Minn APS   Humphreys, R. M. 1993, in Ref. 70; see also Odewahn, S. C, Stockwell, E. B., Pennington, R. L., Humphreys, R. M., and Zumach, W. A. 1992 Astron. J. 103, 318; and Annual Reports for the University of Minnesota 1993, B. A. A. S. 25, 320.

45. STScI dig.   Lasker, B. M. 1992, in Digitized Optical Sky Surveys, ed. H. T. MacGillivray and E. B. Thomson (Dordrecht: Kluwer), p. 87; see also Lasker, B. M. 1993, in Ref. 70. STScI dig. = Space Telescope Science Institute digitization plans.

46. Luyten   Luyten, W. J. 1987, "My First 72 Years of Astronomical Research", (Minneapolis), pp. 26-28.

47. ROE dig.   Working Group on Wide-field Imaging, Newsletter 2, 1992, p. 10. ROE dig. = Royal Observatory Edinburgh digitization plans.

48. Cam. dig.   Irwin, M. J., and McMahon, R. 1992, Working Group on Wide-field Imaging, Newsletter 2, p. 31. Cam. dig. = Cambridge University digitization plans.

49. AAO-SES   Lasker, B. M., and Cannon, R. D. 1989, in Digitized Optical Sky Surveys, ed. C. Jaschek, in Bull. d'Info. du CDS, Number 37, Obs. de Strasbourg, p. 13. AAO-SES = Anglo Australian Observatory Second Epoch Survey.

50. BPM   Luyten, W. J., 1963, BPM = Bruce Proper Motion Survey, University of Minnesota, Minneapolis. Mostly Southern Hemisphere, but includes some in the north.

51. Lowell N   Giclas, H. L., Burnham Jr., R., and Thomas, N. G., 1971, Lowell N = Lowell Proper Motion Survey, Northern Hemisphere Catalogue, Lowell Obs., Flagstaff.

52. Lowell S   Giclas, H. L., Burnham Jr., R., and Thomas, N. G., 1978, Lowell S = Lowell Proper Motion Survey, Southern Hemisphere Catalogue, Lowell Obs. Bull. No. 164, Vol. VIII, 89.

53. LP   Luyten, W. J., 1963-87, Proper Motion Survey with the Forty-Eight inch Schmidt Telescope, University of Minnesota, Minneapolis; and 1987, "My First 72 Years of Astronomical Research", (Minneapolis), pp. 26-28. LP = Luyten Palomar Proper Motion Survey. Mostly Northern Hemisphere, but includes some in the south.

54. Calan   Wroblewski, H., and Torres, C., 1992, Astron. Astrophys. Suppl. Ser. 78, 231, 1989; 83, 317, 1990; 91, 129, 1991; 92, 449, 1992. Calan = Cerro Calan Observatory, Univ. of Santiago, Chile.

55. NLTT   Luyten, W. J., 1979, 80, NLTT = New Luyten Catalogue of Stars with Proper Motions Larger than Two Tenths of an Arcsecond, University of Minnesota, Minneapolis.

56. LHS Cat.   Luyten, W. J., 1979, LHS = Luyten Half-Second Catalogue, University of Minnesota, Minneapolis.

57. LHS Atlas   Luyten, W. J., and Albers, H. 1979, Luyten Half-Second Atlas, University of Minnesota, Minneapolis. Finding charts for stars without published charts.

60. BSC V   Hoffleit, E. D., and Warren Jr., W. H., 1993, BSC V = The Bright Star Catalogue, Fifth Revised Edition, NASA Astronomical Data Center, Greenbelt; See also Hoffleit, E. D. (with the collaboration of Jaschek, C.) 1982, The Bright Star Catalogue, Fourth Revised Edition, Yale Univ. Obs., New Haven, (and Ref. 67).

61. BSC Supp.   Hoffleit, E. D., Saladyga, M., and Wlasuk, P., 1983, A Supplement to the Bright Star Catalogue, Yale Univ. Obs., New Haven, (and Ref. 67).

62. CNS 3   Jahreiß, H., and Gliese, W. 1994, CNS 3 = The Catalogue of Nearby Stars, 3rd edition, Veröff. Astron. Rechen-Instituts, Heidelberg, (in press); See also Ref. 67 for the 1991 Preliminary Edition.

63. YPC   van Altena, W. F., Lee, J. T., and Hoffleit, E. D., 1993, General Catalogue of Trigonometric Parallaxes, Fourth Edition, Yale Univ. Obs., New Haven; See also Ref. 67 for the 1991 Preliminary Edition, and van Altena, et al. for a discussion of the system of the Catalogue in Ref. 70 and Ref. 71. YPC = Yale Parallax Catalogue.

64. WDS   Worley, C. E., and Douglass, G. G., 1984, Washington Catalog of Visual Double Stars, U. S. Naval Obs., Washington, D. C.; new edition scheduled for 1994. WDS = Washington Double Stars.

65. CHARA   McAlister, H. A., and Hartkopf, W. I., 1988, Second Catalog of Interferometric Measurements of Binary Stars, CHARA = Center for High Angular Resolution Astronomy, Georgia State Univ., Atlanta.

66. Orbits   Worley, C. E., and Heintz, W. D., 1983, Fourth Catalog of Orbits of Visual Binary Stars, Publ. U. S. Naval Obs. (2) 24, Part VII.

67. ADC CD   Brotzman, L. E., Gessner, S. E., Mead, J. M., and Van Steenberg, M. E., 1991, ADC CD = Astronomical Data Center CD-ROM Selected Astronomical Catalogs, Volume I, NASA Goddard Space Flight Center, Greenbelt.

68. IAU XXIB   Transactions of the IAU XXIB 1991, edited by J. Bergeron, (Dordrecht: Kluwer).

69. IAU 141   *IAU Symposium* **141**, "Inertial Coordinate System on the Sky", eds. J. H. Lieske and V. K. Abalakin, (Dordrecht: Kluwer).

70. Swarthmore Workshop on Databases for Galactic Structure, held at Swarthmore College in May 1993, edited by A. G. Davis Philip, Bernard Hauck and Arthur R. Upgren, L. Davis Press, Schenectady, (in press).

71. Cambridge   Galactic and Solar System Optical Astrometry: Observation and Application, held at Robinson College, Cambridge University in June 1993, edited by L. V. Morrison, Kluwer Press, Dordrecht, (in press).

Astrometric Data and References for Open Clusters.
Prepared by Imants Platais, Department of Astronomy, Yale University.

The most extensive compilation of proper motion studies in and around open clusters is that of F. van Leeuwen (IAU Symp. 113 "Dynamics of Star Clusters" eds. J. Goodman and P. Hut (1985), p.579.). The following Table represents a continuation of that list in a similar format. Absolute proper motions for 61 open cluster are compiled by H. van Schewick (Veroeff. Astron. Inst. Bonn (1971) No. 84, 1). It must be pointed out that the absolute proper motions may be affected by errors in the correction to absolute proper motion. More reliable absolute proper motions for 241 open clusters will be available after completion of the Hipparcos mission (The Hipparcos Input Catalogue (1992) ESA SP-1136, vol. 7, Annex 3). Detailed and comprehensive information on individual stars in about 500 open clusters is available from the database for stars in open clusters (Mermilliod, J.-C. (1992) Bull. Inform. CDS 40, 115). The columns in the following table list: a number which refers to a reference (**Ref. No.**) immediately following this table, the year of publication (**Publ. Yr.**) of the paper, the **Cluster Name**, the **Field Dimensions** studied in arc-minutes, the **Limiting Magnitude**, the number of stars (**No. Stars**), the number of plates used in the proper motion study (**No. Plates**), the **Scale** of the plates in arc-sec/mm, the approximate **Epoch Difference** in years between the first and second epoch plates, the precision of the proper motions in mas/yr (**Acc**), and the **Maximum Probability** of membership (from 0% to 100%) based on the proper motion study (generally speaking, the larger the number, the more reliable the probabilities of membership should be).

Open Clusters

Ref No.	Publ. Yr	Cluster Name	Field Dimensions	Limiting Mag.	No. Stars	No. Plates	Scale "/mm	<Epoch Diff.>	Acc. mas/yr	Max. Prob.
1	91	NGC 225	30 x 30	16.1	289	7+	60	80	0.6	96
2	89	NGC 752	60 x 50	15.0	457	17	10	67	0.6	99
3	91	NGC 752	110	15.4	1777	11	60	70	0.8	99
4	83	h&chi Per	100	15.5	5527	4+	60	80	2.1	-
5	93	NGC 1039	55 x 40	15.0	354	16	21	55	1.0	90
6	85	Alpha Per	75 x 75	17.	4000	3+	67	25	8.	-
7	92	Alpha Per	360 x 360	19.	779	11+	55	?	4.	-
2	89	NGC 1342	60 x 50	14.5	275	8	10	70	0.4	99
8	85	Pleiades	180 x 180	14.5	1850	166	40	80	0.15	
9	88	Pleiades	150 x 120	14.0	344	6+	30	70	1.2	-
10	91	Pleiades	240 x 240	18.	809	10	55	25	1.0	85
11	91	Pleiades	90 x 90	15.	458	3+	51	33	3.0	-
38	92	Pleiades	175 x 200	18.0	11000	8	51	30	3.0	-
2	89	NGC 1528	45 x 35	14.5	623	21	10	56	0.4	99
39	92	Hyades	650 x 650	19.	450000	24	67	35	6.	-
2	89	NGC 1647	50 x 45	15.0	331	37	10	71	0.4	99

Open Clusters (continued).

Ref No.	Publ. Yr	Cluster Name	Field Dimensions	Limiting Mag.	No. Stars	No. Plates	Scale "/mm	<Epoch Diff.>	Acc. mas/yr	Max. Prob.
37	90	NGC 1647	60 x 60	14.0	87	16	60	90	2.0	-
12	83	NGC 1817	100 x 100	?	752		30	55	3.2	74
13	88	Orion	30	12.5	73	138	10	75	0.15	99
14	88	Orion	30	16.0	1053	29	13	25	1.0:	99
15	89	Orion	150 x 150	14.0	630	85	98	70	0.5	?
16	86	NGC 2168	30	14.7	526	46	10	75	0.3	99
2	89	NGC 2281	55 x 45	15.5	1109	34	10	72	0.5	99
17	90	NGC 2286	100 x 100	15.0	2284	8	30	50	2.0	70
18	87	NGC 2287	50 x 35	12.0	154	12	21	55	1.0	94
19	86	NGC 2301	50	?	190	2	60	75	?	-
20	86	NGC 2324	50	?	98	2	60	70	?	-
21	91	Praesepe	90 x 90	16.2	257	7+	30		0.2	-
22	91	Praesepe	240 x 240	18.	765	9	55	41	2.0	99
23	84	Tr 10	65 x 65	13.0	979	24	60	80	?	-
24	84	NGC 2682	130 x 130	15.8	1068	10	60	60	2.0	-
25	89	NGC 2682	34 x 44	15.5	663	44	10	60	0.2	99
26	88	Upgren 1	30	13.0	19	4	15	20	1.5	-
27	91	IC 4665	60 x 45	14.5	391	20+	10	50	1.4	70
28	86	NGC 6611	75 x 65	15.0	233	15	40	65	1.4	94
2	89	IC 4725	35 x 30	14.5	533	14	10	50	0.6	99
29	91	NGC 6939	18	14.6	146	4+	90	30	3.0	-
30	86	2128+485	8	15.2	47	11	60	70	1.8	-
31	84	NGC 7092	110	16.7	7931	11	60	70	1.5	98
32	92	NGC 7092	110	13.5	654	6+	90	30	2.4	-
33	87	Tr 37	90	15.0	1387	17	10	35	2.0	94
34	91	Tr 37	150 x 150	14.5	826	8	90	30	4.5	-
35	83	NGC 7160	95 x 90	16.	1342	18	60	70	3.0?	-
36	91	NGC 7209	120	15.6	4290	12	60	80	0.6	99

References for the Table of Open Cluster Proper Motion Investigations.

1. Lattanzi M.G., Massone G., Munari U. 1991, Astron. J. 102, 177.
2. Francic S. 1989, Astron. J. 98, 888.
3. Platais I. 1991, Astron. Astrophys. Suppl. 87, 69.
4. Muminov M. 1983, Bull. Inform. CDS 24, 95.
5. Ianna P.A., Schlemmer D.M. 1993, Astron. J. 105, 209.
6. Stauffer J.R., Hartmann L.W., Burham J.N. et al 1985, ApJ 289, 247.
7. Prosser C.F. 1992, Astron. J. 103, 488.
8. van Leeuwen F. 1985, IAU Symp. 113, 477.
9. Zhou X.S., Shu Y.-F., Jiang J.-Y. 1988, Publ. Purple Mount Obs. 7, 330.
10. Stauffer J., Klemola A., Prosser C.F. et al 1991, Astron. J. 101, 980.
11. Schilbach E. 1991, Astron. Nachr. 312, 197.
12. Tian K., Ying M., Jin J., Xu Z. 1983, Ann. Shanghai Obs. 5, 136.
13. van Altena W.F., Lee J.T., Lee J.-F et al 1988, Astron. J. 95, 1744.
14. Jones B.F., Walker M.F. 1988, Astron. J. 95, 1755
15. McNamara B.J., Hack W.J., Olson R.W. et al 1989, Astron. J. 97, 1427.
16. McNamara B.J., Sekiguchi K. 1986, Astron. J. 91, 557.
17. Zhao J.-L., He Y.-P., Tian K.-P. 1990, Chin. Astron. Astroph. 14, 346.
18. Ianna P.A., Adler D.S., Faudree E.F. 1987, Astron. J. 93, 347.
19. Aiad A. 1986, Bull. Fac. Sci. Univ. Cairo 54, 729.
20. Aiad A. 1986, Bull. Fac. Sci. Univ. Cairo 54, 753.
21. Wang. J.-J., Jiang P.-F. 1991, Acta Astron. Sinica 32, 176.
22. Jones B.F., Stauffer J.R. 1991, Astron. J. 102, 1080.
23. Stock J. 1984, Rev. Mex. Astron. Astrofis. 9,127.
24. Frolov. V.N. 1984, Izvest. Pulkovo 202, 105.
25. Girard T.M., Grundy W.M., Lopez C.E. et al 1989, Astron. J. 98, 227.
26. Gatewood G, de Jonge J.K., Castelaz M. et al 1988, ApJ 332, 917.
27. Prosser C.F. 1991, PhD Dissertation, University of California, 228 pp.
28. Tucholke H.-J., Geffert M., The P.S. 1986, Astron. Astrophys. Suppl. 66, 311.
29. Glushkova E.V., Rastorguev A.S. 1991, Sov. Astron. Lett. 17, 64.
30. Platais I.K. 1986, Nauchnye Inf. 61, 89.
31. Platais I.K. 1984, Sov. Astron. Lett. 10, 84.
32. Glushkova E.V. 1992, Sov. Astron. 35, 466.
33. Marschall L.A., van Altena W.F. 1987, Astron. J. 94, 71.
34. Glushkova E.V. 1991, Sov. Astron. Lett. 17, 218.
35. De Graeve E. 1983, Vatican Obs. Publ. 2, 31.
36. Platais I. 1991, Astron. Astrophys. Suppl. 87, 577.
37. Geffert, M., Guibert, J., Ducourant, C. 1990, Colloque. A. Danjon, p. 215.
38. Schilbach, E., Guibert, J., Geffert, M., Hirtes, 1992, IAU Symp. 156 in press,.
39. Reid, N. 1992, Mon. Not. R. Astr. Soc. 257, 257.

Astrometric Data and References for Globular Clusters and Dwarf Spheroidal Galaxies.
Prepared by Kyle M. Cudworth, Yerkes Observatory, University of Chicago.

This table compiles references to papers published 1960-1992, with a few important earlier papers and a few 1993 papers known to be in preparation. The clusters are listed in order of right ascension, and references for each cluster are listed in chronological order. The two studies of dwarf spheroidal galaxies (dSph) are included at the end of this table since these systems have traditionally (though not necessarily correctly) been included with globular clusters in stellar population discussions. The columns list the commonly used **Cluster Names**, the number of stars (**NS**), the approximate limiting magnitude (**m**) for those studies that derive proper motions for individual stars that may be useful in the context of cluster membership, and the absolute proper motions and hence space velocities (**SV**) of the clusters, though such derivations also appear in some of the other papers. The codes **C&H** and **B** refers to the papers by Cudworth and Hanson in AJ 105, 168, 1993, and Brosche, P., Tucholke, H.-J., Klemola, A. R., Ninkovic, S., Geffert, M., and Doerenkamp, P. 1991, AJ, 102, 2022 in which numerous space velocities are determined; **y** indicates that the space velocity has been determined for this cluster. Readers should be cautious, however, since in some cases the proper motion precision is barely adequate to make a preliminary separation of field stars from cluster members. Several papers that discuss subsets of the data from larger papers have not been included here. References without NS and m values listed are usually derivations of absolute proper motions and hence space velocities of the clusters.

Globular Clusters and Dwarf Spheroidal Galaxies.

Cluster Names		NS	m	SV	References
NGC 104	47 Tuc	3076	17	C&H	Tucholke, H.-J. 1992, A&AS, 93, 293.
NGC 288		7000	22.5	y	Guo, X., Girard, T. M., van Altena, W. F., and López, C. E. 1993, AJ 105, 2182.
NGC 362		490	17	B	Tucholke, H.-J. 1992, A&AS, 93, 311.
NGC 4147		42	17.3	B	Brosche, P., Geffert, M., Klemola, A. R., and Ninkovic, S. 1985, AJ, 90, 2033.
NGC 5139	ω Cen				Murray, C. A., Jones , D. H. P., and Candy, M. P. 1965, ROB No. 100.
		4525	16		Woolley, R. v. d. R. 1966, ROA No. 2
		7000	16		Reijns, et al. 1993, in ASP Conf. Ser. "Dynamics of Globular Clusters"
NGC 5024	M53	450	15.5	y	Geffert, M., Forner, C., and Hiessen, M. 1993, in VIII Internat. Colloq. Geodet. Astrom., Dresden, Germany.
NGC 5272	M3			C&H	Zhukov, L. V. 1966 Comm. Pulkova Obs. 24, no. 181, 159
		764	16		Zhukov, L. V. 1969 Sov. Astron. AJ 13, 306
		764	16		Zhukov, L. V. 1971, Tr. Glav. Ast. Obs. Pulkova, 78, 160
		296	14.5		Spasova 1977, Astrofiz. Issled. NRB, 2, 46
		270	16		Cudworth, K. M. 1979, AJ, 84, 1312
					Scholz, R.-D. 1990, in IAU Symp 141, p. 451
				y	Scholz, R. D., Odenkirchen, M., & Irwin, M. J. 1993, MNRAS (submitted).
NGC 5466				B	Brosche, P., and Geffert, M. 1982, Naturwissenschaften, 69, 236
		10	15.5		Brosche, P., and Geffert, M. 1983, A&A, 127, 415
Pal 5		590	21		Cudworth, K. M., Schweitzer, A. and Majewski, S. R. 1993, AJ, (in prep).

Globular Clusters and Dwarf Spheroidal Galaxies (continued)

Cluster Names		NS	m	SV	References
NGC 5904	M5	1141	16	C&H	Zhukov, L. V. 1969 Sov. Astron. AJ 13, 306
		1141	16		Zhukov, L. V. 1971, Tr. Glav. Ast. Obs. Pulkova, 78, 160
		320	15.5		Cudworth, K. M. 1979, AJ, 84, 1866
		322	14.5		Spasova, N. M., and Mikhnevski 1981, Astrofiz. Issled. NRB, 3, 70 Rees, R. 1993, AJ, (in prep)
		550	15.5	y	Geffert, M., Forner, C., and Hiessen, M. 1993, in VIII Internat. Colloq. Geodet. Astrom., Dresden, Germany.
NGC 6121	M4	530	16	C&H	Cudworth, K. M., and Rees, R. 1990, AJ, 99, 1491
NGC 6171	M107	400	17	C&H	Cudworth, K. M., Smetanka, J. J., and Majewski, S. R. 1992, AJ 103, 1252
NGC 6205	M13	412	15.3	C&H	Brown, A. 1955, ApJ, 122, 146
		156	14.5		Pantelyeeva, L. P. 1965, Comm. Sternberg Ast. Inst. no. 140, 58
		923	16.5		Kadla, Z. I. 1966, Comm. Pulkova Obs., 24, no. 181, 93
		334	14.5		Spasova, N. M. 1977, Astrofiz. Issled. NRB, 2, 34.
		440	15.5		Cudworth, K. M., and Monet, D. G. 1979, AJ, 84, 774
		730	16.5		Cudworth, K. M. 1979, AJ, 84, 1005
NGC 6218	M12			B	Geffert, M., Tucholke, H.-J., Georgiev, Ts., LeCampion, J.-F. 1991, A&AS 91, 487
NGC 6341	M92	117	14.5	C&H	Spasova, N. M. 1977, Astrofiz. Issled. NRB,
		220	15.5		Cudworth, K. M. 1976, AJ, 81, 975
		365	16		Rees, R. 1992, AJ 103, 1573
NGC 6397		600	16	C&H	Cudworth, K. M., and Benensohn, J. 1993, AJ, (in prep)
NGC 6522		250	18		Clube, S. V. M. 1966, ROB no. 111
NGC 6626	M28	250	16	C&H	Rees , R, and Cudworth, K. M. 1991, AJ, 102, 152
NGC 6656	M22	672	15	C&H	Cudworth, K. M. 1986, AJ, 92, 348
	(pl. neb.)				Cudworth, K. M. 1990, AJ, 99, 1863
NGC 6712		600	18	C&H	Cudworth, K. M. 1988, AJ, 96, 105
NGC 6779	M56	540	15	C&H	Rishel, B. E., Sanders, W. L., and Schröder, R. 1981, A&AS, 45, 443
NGC 6838	M71	400	15		Sanders, W. L. 1971, A&A, 15, 173
		360	16		Cudworth, K. M. 1985, AJ, 90, 65
NGC 7078	M15	290	15	C&H	Brown, A. 1951, ApJ, 113, 344
		210	15		Cudworth, K. M. 1976, AJ, 81, 519
		277	14.5		Spasova, N. M. 1981, Astrofiz. Issled. NRB, 3, 63
		863	15		Le Campion, J. -F. Geffert, M., Dulou, M. R., and Colin, J. 1992, A&AS, 95, 233
		863	15	y	Geffert, M., Colin, J., Le Campion, J.-F., & Odenkirchen, M. 1993, AJ (submitted)
NGC 7089	M2	300	16	C&H	Cudworth, K. M., and Rauscher, B. J. 1988, AJ, 93, 856
Ursa Minor	Dw. Sph.	500	20.5		Cudworth, K. M., Olszewski , E. W., and Schommer, R. A. 1986, AJ, 92, 766
Draco	Dw. Sph.	525	21		Stetson, P. B. 1980, AJ 85, 387

Radio Astrometry References
Prepared by J. Russell and K. Johnston, U. S. Naval Research Laboratory

Ref.	Project/Catalog	Ns	s.e.(")	Data/Description	Band	RA Zero Pt. Def.	Dec. Lim	Notes
90-5	Radio/Optical tie	400+	0.001	Astrometry + Geodesy	S/X	27 QSO's (FK5)	-82,+85	1*
96	JPL	282	0.001	Geodesy (DSN+)	S/X	0851+202+CTD20	-45,+85	2*
97	IERS	422	0.001	Compilation	S/X	"Primary sources"	-82,+85	3*
98-9	VLA calibration	595	var.	Phase reference calibration	S/X, L/C	15 Ref. souces	-46,+86	4
100	MERLIN Survey	800	0.012	VLA snapshots	X	Relative to Rad./Opt.	+35,+75	5*

The column labeled **Ref.** refers to entries in the Bibliography, which contains detailed references for recent publications in each project. Machine readable versions of all of the catalogs may be obtained from the authors and/or the Astronomical Data Center at Goddard Space Flight Center and at Data Centers in other countries. **Ns** is the number of sources in the catalog, **s.e.(")** is a representative standard error for the majority of radio positions in arc seconds, **Band** defines the wavelength of the observations (L=20cm, C=6cm, X=3.6cm, S=13cm), **RA Zero Pt. Def.** lists the method used to establish the zero point of the Right Ascension, **Dec. Lim** gives the declination range of the list, and the **Notes** refer to an entry in the following list.

Notes:

* on-going VLBI projects; check for periodic updates with similar publications/authors.

1) Effort to tie radio and optical reference frames, coordinated by U. S. Navy with many cooperating agencies and observatories, uses available astrometric and geodetic data. New and updated source positions published at least annually.
2) Geodetic data from Deep Space Network (DSN) for reference and time keeping. Annual updates in JPL Technical reports.
3) International Earth Rotation Service (IERS) combines geodetic data from NASA/CDP (Crustal Dynamics Project), Jet Propulsion Lab. (JPL), Naval Research Lab. (NRL), U. S. Naval Obs. (USNO) and the Nat. Oceanic & Atmospheric Admin. (NOAA). Annual reports published about mid-year.
4) The list of calibrators for the Very Large Array (VLA) contains 814 sources of which 595 are designated as position calibrators. Of these, 294, mostly from JPL, are recommended if errors <0.01 arcsec are required in absolute position. Many others based on Perley list of 404 sources with formal s.e. 0.05 arcsec.
5) Survey for phase reference sources for MERLIN array, additional declination zones planned. Positions based on VLA snapshots referred to 63 sources from Radio/Optical reference frame.

Radio Astrometry Bibliography

90. Project description: Johnston, K. J., Russell, J., de Vegt, C., Hughes, J., Jauncey, D., White, G., and Nicolson, G. (1988) in "The Impact of VLBI on Astrophysics and Geophysics," IAU Symp. 129, M. Reid and J. Moran, eds., (Reidel, Dordrecht), 317.
91. Paper 1: Ma, C., Shaffer, D. B., de Vegt, C., Johnston, K. J., Russell, J. L., (1990) Astron. J. 99, 1284.
92. Paper 2: Russell, J. L., Johnston, K. J., Ma, C., Shaffer, D. B., de Vegt, C., (1991), Astron. J., 101, 2266-2273.
93. Paper 3: Russell, J. L., Jauncey, D. L., Reynolds, J. E., White, G., Nothnagel, A., Nicolson, G., Ma, C., Kingham, K., Johnston, K. J., Malin, D., (1992) Astron. J., 103, 2090.
94. Paper 4: Fey, A., Russell, J. L., Ma, C., Johnston, K. J., Archinal, B. A., Carter, M. S., Holdenreid, E., Yao, Z., (1992) Astron. J. 104, 891.
95. In preparation: Russell et al., Fey et al., Reynolds et al.
96. Steppe, J.A., Gross, R.S., Sovers, O.J. (1992) JPL Geodesy and Geophysics Preprint No. 217, Jet Propulsion Laboratory, Pasadena CA.
97. IERS Annual Report, publ annually by Central Bureau of IERS -- Observatoire de Paris; 61, avenue de l'Observatoire; F-75014 Paris, France.(Most recent report for 1991 data was published July 1992.)
98. VLA Calibration Manual (1990) distributed by NRAO, P.O. Box O, Socorro NM 87801.
99. Perley, R. A. (1982) Astron. J. 87, 859-880.
100. Patnaik, A. R., Browne, I. W. A., Wilkinson, P. N., Wrobel, J. M. (1992) Mon. Not. R. Astr. Soc. 254, 655-676.

INDICES

If a page number is underlined in the Name Index it indicates the name of an author of a paper. An underlined page number in the Object or Subject Indices indicates that the object or subject was mentioned in the title of a paper. Underlined page numbers of institutions are the institutions of authors of papers. Subjects that are mentioned in major headings may be listed in the index but their page numbers are not underlined.

OBJECT INDEX

SUBJECT INDEX

Other Publications in Astronomy and Mathematics

THE EVOLUTION OF POPULATION II STARS (1972), A. G. D. Philip, ed., Dudley Observatory Report No. 4.

MULTICOLOR PHOTOMETRY AND THE THEORETICAL HR DIAGRAM (1975), A. G. D. Philip and D. S. Hayes, eds., Dudley Observatory Report No. 9.

UBV COLOR-MAGNITUDE DIAGRAMS OF GALACTIC GLOBULAR CLUSTERS (1976), A. G. D. Philip, M. F. Cullen and R. E. White, Dudley Observatory Report No. 11.

AN ANALYSIS OF THE HAUCK-MERMILLIOD CATALOGUE OF HOMOGENEOUS FOUR-COLOR DATA (1976), A. G. D. Philip, T. M. Miller and L. J. Relyea, Dudley Observatory Report No. 12.

GALACTIC STRUCTURE IN THE DIRECTION OF THE POLAR CAPS (1977), M. F. McCarthy and A. G. D. Philip, eds., in Highlights of Astronomy, Vol 4, Reidel, Dordrecht.

IN MEMORY OF HENRY NORRIS RUSSELL (1977), A. G. D. Philip and D. H. DeVorkin, eds., Dudley Observatory Report No. 13.

IAU SYMPOSIUM NO. 80, THE HR DIAGRAM (1978), A. G. D. Philip and D. S. Hayes, eds., Reidel, Dordrecht.

IAU COLLOQUIUM NO. 47, SPECTRAL CLASSIFICATION OF THE FUTURE (1979), M. F. McCarthy, A. G. D. Philip and G. V. Coyne, eds., Vatican Observatory.

PROBLEMS OF CALIBRATION OF MULTICOLOR PHOTOMETRIC SYSTEMS (1979), A. G. D. Philip, ed., Dudley Observatory Report No. 14.

X-RAY SYMPOSIUM 1981 (1981), A. G. D. Philip, ed., L. Davis Press.

IAU COLLOQUIUM NO. 68, ASTROPHYSICAL PARAMETERS FOR GLOBULAR CLUSTERS (1981), A. G. D. Philip and D. S. Hayes, eds., L. Davis Press.

A DEEP OBJECTIVE PRISM SURVEY OF THE LARGE MAGELLANIC CLOUD FOR OB AND SUPERGIANT STARS. PART I. (1983), A. G. D. Philip and N. Sanduleak, L. Davis Press.

IAU COLLOQUIUM NO. 76, THE NEARBY STARS AND THE STELLAR LUMINOSITY FUNCTION (1983), A. G. D. Philip and A. R. Upgren, eds., L. Davis Press.

IAU SYMPOSIUM NO. 111, CALIBRATION OF FUNDAMENTAL STELLAR QUANTITIES (1985), D. S. Hayes, L. Pasinetti and A. G. D. Philip, eds., Reidel, Dordrecht.

IAU COLLOQUIUM NO. 88, STELLAR RADIAL VELOCITIES (1985), A. G. D. Philip and D. W. Latham, eds., L. Davis Press.

HORIZONTAL-BRANCH AND UV-BRIGHT STARS (1985), A. G. D. Philip, ed., L. Davis Press.

SPECTROSCOPIC AND PHOTOMETRIC CLASSIFICATION OF POPULATION II STARS (1986), A. G. D. Philip, ed., L. Davis Press.

STANDARD STARS (1986), A. G. D. Philip, ed., in Highlights of Astronomy, Vol 7.

IAU COLLOQUIUM NO. 95, THE SECOND CONFERENCE ON FAINT BLUE STARS (1987), A. G. D. Philip, D. S. Hayes and J. W. Liebert, eds., L. Davis Press.

IAU SYMPOSIUM NO. 126, GLOBULAR CLUSTER SYSTEMS IN GALAXIES (1988), J. E. Grindlay and A. G. D. Philip, eds., Reidel, Dordrecht.

NEW DIRECTIONS IN SPECTROPHOTOMETRY (1988), A. G. D. Philip, D. S. Hayes and S. J. Adelman, eds, L. Davis Press.

CALIBRATION OF STELLAR AGES (1988), A. G. D. Philip, ed., L. Davis Press.

STAR CATALOGUES: A CENTENNIAL TRIBUTE TO A. N. VYSSOTSKY (1989), A. G. D. Philip and A. R. Upgren, eds., L. Davis Press.

THE GRAVITATIONAL FORCE PERPENDICULAR TO THE GALACTIC PLANE (1989), A. G. D. Philip and P. K. Lu, eds., L. Davis Press.

CCDs IN ASTRONOMY. II. NEW METHODS AND APPLICATIONS OF CCD TECHNOLOGY (1990) A. G. D. Philip, D. S. Hayes and S. J. Adelman, eds, L. Davis Press.

PRECISION PHOTOMETRY: ASTROPHYSICS OF THE GALAXY (1991), A. G. D. Philip, A. R. Upgren and K. A. Janes, eds., L. Davis Press.

Mm, Vol I, SERIES ON FRACTALS, MIDGETS ON THE SPIKE (1991), A. G. D. Philip, A. Robucci, M. Frame and K. W. Philip, L. Davis Press.

OBJECTIVE-PRISM AND OTHER SURVEYS. A MEETING IN MEMORY OF NICHOLAS SANDULEAK (1991), A. G. D. Philip and A. R. Upgren, eds., L. Davis Press.

NEW YORK STATE ASTRONOMY (1992), A. G. D. Philip, ed., L. Davis Press.